Lester Frank Ward

Synopsis of the flora of the Laramie group

Lester Frank Ward

Synopsis of the flora of the Laramie group

ISBN/EAN: 9783337268626

Printed in Europe, USA, Canada, Australia, Japan

Cover: Foto ©berggeist007 / pixelio.de

More available books at **www.hansebooks.com**

SYNOPSIS

OF THE

FLORA OF THE LARAMIE GROUP.

BY

LESTER F. WARD.

CONTENTS.

ILLUSTRATIONS.

SYNOPSIS OF THE FLORA OF THE LARAMIE GROUP.

INTRODUCTION.

The object of this paper is twofold: first, to offer, as its title implies, a synopsis, or condensed account, of the flora of the Laramie group, as that formation is now understood; and, secondly, to give a few illustrations of this flora from new material or from material more ample and abundant than has heretofore existed.

Mr. Leo Lesquereux, in his "Tertiary Flora,"[1] describes a large number of plants belonging to this group, but he here argues for the Tertiary age of these plants and regards the group as Eocene; he therefore makes no attempt to keep them separate from those derived from higher and still acknowledged Tertiary beds. In his last work, on "The Cretaceous and Tertiary Floras,"[2] he attempts to introduce a "table of distribution" of the plants of the Laramie group, but in doing so he fails to recognize the Fort Union forms as belonging to that group, although the identity of the two groups had been admitted by Dr. Hayden in his annual reports and was reasserted in his letter transmitting Mr. Lesquereux's "Tertiary Flora" to the Secretary of the Interior for publication. He preferred to accept the view of Mr. Clarence King (who admitted that he had not visited the Fort Union beds), as expressed in his Report of the Geological Exploration of the Fortieth Parallel, Volume I, pp. 353, 354, and which rested upon the determinations by Dr. Newberry of certain vegetable remains of Miocene type. Mr. King believed this formation to be equivalent to the White River Miocene, and Dr. Newberry referred all his Fort Union plants to the Miocene. The only localities which he admits as constituting the plant beds of the Laramie group known at that date are those of Colorado, the Raton Mountains, Placière, Henry's Fork, Barrel Springs, Fort Ellis, Spring Cañon, Black Buttes, Point of Rocks, and Yellowstone Lake. This excludes Carbon and Evanston, which I shall also embrace in the Laramie, and there are several other localities from which fossil plants have been obtained that belong with little doubt to the same great system.

[1] Contributions to the Fossil Flora of the Western Territories, Part II. The Tertiary Flora. By Leo Lesquereux. Report of the United States Geological Survey of the Territories, F. V. Hayden, United States geologist-in-charge. Vol. VII. Washington, 1878.

[2] Contributions to the Fossil Flora of the Western Territories, Part III. The Cretaceous and Tertiary Floras. By Leo Lesquereux. Report of the United States Geological Survey of the Territories, F. V. Hayden, geologist-in-charge. Vol. VIII. Washington, 1883.

HISTORICAL REVIEW OF OPINION.

The history of the Laramie group, as now understood, is a long one, and the literature is scattered through a series of reports in a manner very perplexing to any one who desires to gain a comprehensive knowledge of it. From the circumstance that at nearly all places where it has been recognized it consists to a greater or less extent of deposits of lignite or coal, this condition was for a time inseparably associated with it to such an extent that there was a disposition to regard all the lignitic deposits of the West as belonging to the same geologic formation; but when this had been disproved by the discovery of extensive beds of coal in the middle Cretaceous, the reaction against this view carried many too far, and resulted in the quite general belief that the lignite beds of the Upper Missouri River were of widely different age from those of Colorado and Wyoming. Even Mr. King, who correlated all the beds along the 40th parallel, and first gave them the name of "Laramie group," still denied the identity of the Fort Union beds with them, and as late as 1878 regarded these as Miocene and the equivalent of those of the White River. It is remarkable that he should have expressed such an opinion in so prominent a place as his final report (Report of the Geological Survey of the Fortieth Parallel, Vol. I, p. 353), while admitting that he had not personally examined this region.

The northern portion of the extensive area now embraced under the name Laramie group was the first to attract attention. It was natural that the earliest transcontinental voyages should follow the largest water-ways, and notwithstanding the extremely slow development of the Upper Missouri River region we find that its exploration was begun in the first decade of the century by parties provided with appliances for scientific observation and has been continued at intervals ever since. Leaving the merely geographical aspects out of the account, we find that the coal beds attracted the attention of Lewis and Clarke in 1803 and of every subsequent expedition down to the epoch of true geologic investigation, which dates from the commencement of the protracted researches of Messrs. Meek and Hayden in the year 1854, the earliest publications of which are contained in Volume VIII of the Proceedings of the Philadelphia Academy of Sciences, 1856. The investigations of Harris and Audubon in 1844[1] added scarcely anything to the knowledge of the geological age of these regions. As much might be said of the explorations of Frémont, who observed the lignite beds of Wyoming in 1842, and of the expedition of General Emory who noted those of Eastern New Mexico in 1848. But the large collections brought by Hayden from Nebraska and the Upper Missouri and Yellowstone regions in 1854 furnished the data for profitable scientific in-

---

[1] Proceedings of the Academy of Natural Sciences, Philadelphia, Vol. II, 1845, pp. 235-240.

vestigation, which they soon received at the competent hands of Messrs.
Meek and Leidy. In the first of the papers above referred to,[1] in which
all the species described are mentioned as Cretaceous, the authors
remark : " It is worthy of note that some of the species contained in the
collection from the most recent Cretaceous beds of the Upper Missouri
country appear referable to genera which, according to high European
authority, date no farther back than the true chalk, while many of them
are closely analogous to Tertiary forms; so close, indeed, that, had they
not been found associated in the same beds with Ammonites, Scaphites,
and other genera everywhere regarded as having become extinct at the
close of the Cretaceous epoch, we would have considered them Tertiary
species." A section is given, at the top of which 400 to 600 feet of
" Tertiary" are placed, which is described as " beds of clay, sandstone,
lignite, &c., containing remains of vertebrata, and at places vast numbers of plants, with land, fresh-water, and some times marine or estuary
mollusca."

At the next meeting of the Academy, Dr. Joseph Leidy read a paper
in which he described the vertebrate remains which Dr. Hayden had
obtained from the Bad Lands of the Judith River. He is silent as to
the age which these remains indicate until the close of the paper, where
he names a species of Lepidotus in honor of the discoverer, and says :
"This species is named in honor of Dr. Hayden, who collected the remains characterized in this paper ; and which remains, I suspect, indicate the existence of a formation like that of the Wealden of Europe;"
a remark which has since been much quoted in support of the Mesozoic
age of the Judith River beds.

On June 10th of the same year a second paper was presented to the
Academy by Messrs. Meek and Hayden, entitled " Descriptions of new
species of Acephala and Gasteropoda, from the Tertiary formations of
Nebraska Territory, with some general remarks on the Geology of the
country about the sources of the Missouri River."

These " general remarks," which were " based upon the observations
and collections of Dr. Hayden," contain some very interesting statements and certain somewhat remarkable adumbrations of the conclusions to which the latest investigations have led respecting the geology
of this region. The lignitic deposits are regarded as Tertiary, but they
are very clearly distinguished from the fresh-water deposits of the
White River group as well as from the underlying Cretaceous formation.
"Although there can be no doubt," the authors say, " that these deposits
hold a rather low position in the Tertiary system, we have as yet been
able to arrive at no very definite conclusions as to their exact synchronism with any particular minor subdivision of Tertiary, not having
been able to identify any of the mollusca found in them with those of
any well marked geological horizon in other countries. Their general

---

[1] Proceedings of the Academy of Natural Sciences, Philadelphia, Vol. VIII, 1856, p.
63. (Read March 11.)

resemblance to the fossils of the Woolwich and Reading series of English geologists, as well as to those of the great Lignite formations of the southeast of France, would seem to point to the lower Eocene as their position." In view of the fact that eminent geologists with abundant material before them have until very recently regarded the Fort Union group as of Miocene age, this early hint at their lower position seems to deserve mention in passing. On the other hand, the extremes to which certain vertebrate remains from the Judith River beds farther up the Missouri had led paleontologists in the opposite direction were fairly anticipated in this early paper. After commenting upon the facts which prompted Dr. Leidy to liken the Judith River deposits to the Wealden of Europe, the authors add: "Inasmuch, however, as there certainly are some outliers of fresh-water Tertiary in these Bad Lands, we would suggest that it is barely possible these remains may belong to that epoch, though the shells appear to be all distinct species from those found in the Tertiary at all the other localities in this region."

In a subsequent paper, read November 11th of that year and published in the same volume (pp. 265–286), yielding to the weight of authority of the eminent paleontologists who had studied the vertebrate and vegetable remains, these authors, in the section drawn up on page 269, place the yellowish sandstones of the Judith in their lowest member of the Cretaceous (No. 1), along with the darker sandstones of the Big Sioux, now so well known to characterize the Dakota group,[1] while the lignite deposits of the Lower Yellowstone and Fort Union region are put at the top of the Tertiary system and designated as Miocene. In an elaborate paper by Messrs. James Hall and F. B. Meek in the "Memoirs of the American Academy of Arts and Sciences" communicated June 27, 1854,[2] a section is given in which the Cretaceous series is subdivided into five members, corresponding substantially with that published in the Proceedings of the Philadelphia Academy by Messrs. Meek and Hayden (Vol. VIII, 1856, p. 269), as also with that which appeared in the same publication for December, 1861 (Vol. XIII, p. 419), and was reproduced in Hayden's First Annual Report of the United States Geological Survey of the Territories for 1867, where, for the first time, the names by which the groups have since become so widely known were attached. In this earliest section of Meek and Hall the Bad Land formation of the Upper Missouri is placed above the Cretaceous series, and is not subdivided but is designated as "Eocene Tertiary" and assigned a maximum thickness of 250 feet.

On May 26, 1857, Dr. F. V. Hayden laid before the Philadelphia

[1] This view seems to have been maintained by Mr. Meek as late as 1860. See Proceedings of the Academy of Natural Sciences, Philadelphia, Vol. XII (April), 1860, p. 130.

[2] Descriptions of new species of fossils from the Cretaceous formations of Nebraska, &c., Vol. V, 1853, Part II, Art. xvii (extras dated 1856).

Academy a rough geological map of the country bordering on the Missouri River, from the mouth of the Platte to Fort Benton, with explanations.[1] According to this map the "Great Lignitic Tertiary Basin" begins at the mouth of Heart River and extends to near the Muscle Shell. It also stretches back on the Little Missouri to near the base of the Black Hills and on the Yellowstone to the mouth of the Big Horn. He also lays down an extensive "Tertiary" tract lying between the South Fork of the Cheyenne and the Platte and extending east and west from the 100th meridian to Fort Laramie. The Judith River Bad Lands are also treated as Tertiary, the too deep coloring of the map being explained in a foot note on page 110. Of the Great Lignitic deposit he remarks that "the collections of fossils now obtained show most conclusively * * * that it cannot be older than the Miocene period." Of the Judith River basin he says that "the impurity of the lignite forms the most essential lithological difference between this deposit and the Great Lignite basin below Fort Union."

Immediately following this communication in the same volume is a more extended one by Messrs. Meek and Hayden, devoted primarily to the description of new paleontological material from the same general region, but containing an introductory discussion of the geological problems involved. Besides sections of the beds above Fort Clarke, and near the mouth of the Judith, this paper gives a general one for the whole of this country, in which the "Tertiary system" is now classed as Miocene.

The first complete section of the "Tertiary" formations of the West was drawn up by Messrs. Meek and Hayden, and also published in the Proceedings of the Academy of Natural Sciences of Philadelphia, for December, 1861 (Vol. XIII, p. 433). The series is subdivided into the four familiar groups: 1, Fort Union, or Great Lignitic; 2, Wind River; 3, White River; 4, Loup River. We are concerned here only with the first, or lowest member of this series, the so-called Great Lignitic. This is defined as "Beds of clay and sand, with round ferruginous concretions, and numerous beds, seams, and local deposits of lignite; great numbers of dicotyledonous leaves, stems, etc., of the genera Platanus, Acer, Ulmus, Populus, etc., with very large leaves of true fan palms. Also, Helix, Melania, Vivipara, Corbicula, Unio, Ostrea, Potamomya, and scales of Lepidotus, with bones of Trionyx, Emys, Compsemys, Crocodilus, etc.; thickness: 2,000 feet or more; localities: occupies the whole country around Fort Union, extending north into the British possessions to unknown distances; also southward to Fort Clarke. Seen under the White River group on North Platte River above Fort Laramie. Also on west side Wind River Mountains."

Although nothing is said either here or in the more general description which follows of the relation of the Judith River beds to this formation, we learn from a foot note appended to page 417 that the

---

[1] See Proceedings of the Academy of Natural Sciences, Philadelphia, Vol. IX, p. 109.

idea that it could be Jurassic had now been wholly given up by the authors, who had come to regard it as the lower-part of the Fort Union group. This note is as follows: "At the time we published these facts, we were led by the discovery here of fresh-water shells in such a position to think that some estuary deposits of doubtful age near the mouth of the Judith River on the Missouri, from which Dr. Leidy had described some saurian remains resembling Wealden types, might be older than Tertiary. Later examinations, however, have demonstrated that the Judith beds contain an entirely different group of fossils from those found in the rock under consideration, and that they are really of Tertiary age, and hold a position at the base of the Great Lignite series of the Northwest."

In discussing this same section in the First Annual Report of the Geological Survey of the Territories, 1867, Dr. Hayden distinctly classes the Judith River basin with the Fort Union group, and says: "This basin is one of much interest, as it marks the dawn of the Tertiary period in the West by means of the transition from brackish to strictly fresh-water types. It is also remarkable for containing the remains of some curious reptiles and animals, reminding the paleontologist of those of the Wealden of England."

By this time the more southern extension of the coal-bearing beds had begun to receive the attention of geologists, and they had been traced into Wyoming and Colorado and as far south as Raton Pass in New Mexico. Fossil plants had been found at nearly all points, and their testimony was considered the most unanswerable for the Tertiary age of the entire group. Indeed, down to 1868, with the single exception of the alleged Wealden facies of the Judith vertebrates, there was substantial harmony upon this point. The array of names of those who had committed themselves to this view after thorough study of the different kinds of fossils is truly formidable, and there can be no wonder that when their position was at length challenged and the Cretaceous age of this great series asserted the conflict of opinion resulting was sharp and the resistance stubborn. Messrs. Meek, Hayden, Lesquereux, and, as Dr. Hayden states,[1] Leidy, all conceded this. Capt. E. L. Berthoud had studied the formation in Colorado and inclined to take the same view.[2] He says: "Everything that I have so far seen points out that the coal is either Cretaceous or Tertiary, but I believe it to be Tertiary, or of the same age as the coal near Cologne, on the Rhine." In an article contributed by Dr. Hayden to the American Journal of Science for March, 1868 (Vol. XLV, p. 198), he reiterates his views in a form that indicates that thus far they had met with no serious opposition.

The first dissenting voice to this general current of belief seems to have been raised by Dr. John L. LeConte, who had investigated the

[1] Annual Report United States Geological and Geographical Survey of the Territories, 1874, p. 21.

[2] First Annual Report United States Geological and Geographical Survey, 1867, p. 57.

coal and plant bearing beds lying along the Smoky Hill Fork of the
Kansas River. In his report of a survey of this region[1] he gives it as
his opinion that the lignitic strata of this region are older than those
of the Upper Missouri, which he admits to be Miocene (p. 65). He
states that specimens of Inoceramus were found with the coal in Raton
Pass, indicating its Cretaceous age, and then proceeds to adduce reasons for discrediting the evidence furnished by vegetable remains.

The following year (1869) Prof. E. D. Cope, in an exhaustive paper
on the vertebrate paleontology of America, published in the Transactions of the American Philosophical Society (Vol. XIV), in commenting upon *Ischyrosaurus antiquus*, Leidy, from Moreau River, Great Lignitic of Nebraska, speaks of that formation as " perhaps of the Cretaceous age " (p. 40), and with more confidence later on assigns *Hadro-
saurus ? occidentalis*, Leidy, to the "?Cretaceous beds of Nebraska,"
although *Palaeoscincus costatus*, Leidy, is still kept in the " upper Jurassic Bad Lands of Judith River." In the tabular exhibit at the close of
this memoir the first of these species is placed in the Cretaceous column ; the second is also placed in that column, but with an accompanying mark of interrogation, while the third is assigned to the Jurassic
column.

The Third Volume of the United States Geological Exploration of
the 40th Parallel, relating to Mining Industry, bears date 1870, and contains an important chapter (VII) from the pen of Mr. King on the Green
River Coal Basin, in which he maintains that the extensive coal-bearing deposits of this region are chiefly of Cretaceous age, but admits
that the uppermost strata pass into the Tertiary and become fresh-
water beds. He also declares that the true fresh-water Tertiary strata
of the Green River group overlie the coal beds unconformably at all
points. "The fossil life," says Mr. King, " which clearly indicates a
Cretaceous age for the deepest members up to and including the first
two or three important coal beds, from that point gradually changes
with a corresponding alteration of the sediments, indicating a transition
to a fresh-water period. The coal continued to be deposited some time
after the marine fauna had been succeeded by fresh-water types. The
species of fossils are in no case identical with the California Cretaceous
beds, which occupy a similar geological position on the west of the
Sierra Nevada. Their affinities decidedly approach those of the Atlantic
slopes, while the fresh-water species, which are found in connection with
the uppermost coal beds, seem to belong to the early Tertiary period."
And, speaking of the unconformity of strata above referred to, he remarks : " Whatever may be the relations of these beds in other places,
it is absolutely certain that within the region lying between the Green
River and the Wahsatch, and bounded on the south by the Uintah

[1] Notes on the Geology of the Survey for the extension of the Union Pacific Rail-
way, E. D., from the Smoky Hill River, Kansas, to the Rio Grande. By John L.
LeConte, M. D. Philadelphia. February, 1868.

range, there is no single instance of conformity between the coal beds
and the horizontal fresh-water strata above them."

This chapter also contains a list of the fossil invertebrata collected
in that region and named by Mr. Meek, accompanied by an interesting letter explanatory of their geologic significance. The fact that
several species of Inoceramus, and some which seemed referable to
Anchura, were positively credited to the coal series, led Mr. Meek to
speak with the greatest caution as to the age of these rocks; but it is
clear that, but for these facts, coupled with the stratigraphical considerations urged by Mr. King, he would have scarcely hesitated to pronounce it Tertiary. But he lays great stress upon "the fact that these
fossils are all marine types," and says: "From all the facts now known
I can, therefore, scarcely doubt that you are right in referring these
beds to the Cretaceous." A paragraph on page 462 gives his reasons
for this conclusion more in full, together with certain qualifications
which he feels obliged to make, and closes with the remark that the
facts seem to indicate "that these beds belong to one of the very latest
members of the Cretaceous; or, in other words, that they were probably
deposited when the physical conditions favorable to the existence of
those forms of Molluscan life peculiarly characteristic of the Cretaceous
period were drawing to a close or had in part ceased to exist."

Relative to the age of the so-called Bear River estuary beds, Mr.
Meek expressed himself in this communication with still greater reserve.
These beds had been referred by him and Mr. Henry Engelmann to the
Tertiary in 1860, in a communication made by them to Capt. J. H.
Simpson, and published in the Proceedings of the Academy of Natural
Sciences of Philadelphia for April of that year (Vol. XII, p. 130). He
admits, however, that they may be Cretaceous, as they belong to the
lower disturbed system elsewhere regarded as Cretaceous. He says
that some of the fossils described by him from the mouth of the Judith
River "are identical with those found in these Bear River estuary beds,"
expresses doubt that the saurian remains from there were really from
the same horizon, and concludes as follows: "While I am, therefore,
willing to admit that facts may yet be discovered that will warrant the
conclusion that some of these estuary beds, so widely distributed here,
should be included rather in the Cretaceous than in the Tertiary, it
seems to me that such evidence must either come from included *vertebrate* remains or from further discoveries respecting the stratigraphical
position of these beds with relation to other established horizons, since
all the molluscan remains yet known from them (my own opinions are
entirely based on the latter) seem to point to a later origin."

Prof. O. C. Marsh, in giving an account, in the American Journal of
Science for March, 1871, of an expedition conducted by him the previous season through a portion of the Green River Valley and Eastern
Utah, describes the coal deposits met with by the party on Brush Creek
with special reference to their geologic age. He says (p. 195): "As the

age of the coal deposits of the Rocky Mountain region has of late been much discussed, a careful examination was made of the series of strata containing the present bed and their Cretaceous age established beyond a doubt. In a stratum of yellow calcareous shale which overlies the coal series conformably, a thin layer was found full of *Ostrea congesta*, Conrad, a typical Cretaceous fossil; and just above, a new and interesting crinoid, allied apparently to the *Marsupites* of the English Chalk. In the shales directly below the coal bed, cycloidal fish scales and coprolites were abundant; and lower down, remains of turtles of Cretaceous types, and teeth of a Dinosaurian reptile, resembling those of *Megalosaurus*, were also discovered."

The gradual acceptance of the Cretaceous character of the coal-bearing series of the central and southern districts did not thus far shake the opinion of geologists as to the Tertiary age of the Fort Union group. This is reaffirmed in a very positive manner in the Fourth Annual Report of the Geological Survey of the Territories, 1870 (published in 1871), by Dr. J. S. Newberry, who had been long and carefully studying the vegetable remains collected near Fort Union and along the lower Yellowstone, and had already published descriptions of the species.[1] At the time this paper was presented there was no difference of opinion and the evidence of the plants was regarded as simply confirmatory of Meek's conclusions as to the Miocene age of these beds.

Further on in this report (pp. 164, 165) Dr. Hayden discusses the age of the Wyoming coal strata, and says: "So far as we can determine, the coal beds of the Laramie plains are of Eocene age, although the plants are more closely allied to those of the Miocene period of the Old World;" and again: "That there is a connection between all the coal beds of the West I firmly believe, and I am convinced that in due time that relation will be worked out and the links in the chain of evidence joined together. That some of the older beds may be of upper Cretaceous age I am prepared to believe, yet until much clearer light is thrown upon their origin than any we have yet secured I shall regard them as belonging to my transition series, or beds of passage, between the true Cretaceous and the Tertiary."

In the same report Mr. Lesquereux discusses the fossil plants from Raton Pass, collected by Dr. LeConte, whose views have already been stated, as well as those brought in from points along the line of the Union Pacific Railroad and from other parts of the West. He considers them all Tertiary and ranging from the Eocene to the Miocene.

In the corresponding report for 1871, published in 1872, Mr. Lesquereux describes a mass of new material, and from all the data at hand essays a number of important generalizations. As he still regards all the localities in the great coal bearing series of the West as belong-

---

[1] Notes on the Later Extinct Floras of North America, with Descriptions of some New Species of Fossil Plants from the Cretaceous and Tertiary Strata. Annals of the Lyceum of Natural History, New York (April), 1868. (Read April 22, 1867.)

ing to the Tertiary formation, the only point of special interest brought
forth is his attempt to subdivide the American Tertiary into subordi
nate groups based-upon the analogies afforded by their floras with
those of established horizons in Europe and elsewhere. Thus to the
Eocene he refers Raton Pass and Purgatory Cañon, in New Mexico;
Marshall's Mine, in Colorado; Washakie Station and Evanston, in Wyoming; and Spring Cañon, near Fort Ellis, in Montana, as well as Yellowstone Lake, which also belongs to the upper district. To the Lower
Miocene he refers Carbon Station, Junction Station, Medicine Bow,
Rock Creek, and the Washakie group, in Wyoming; and the Fort Union
group, in Montana and Dakota. To the Middle Miocene are referred
Barrel Springs and Muddy Creek, in Wyoming; Henry's Fork of Snake
River; and Elko Station, Nevada. Among the localities the geological
position of which is marked as unknown are the important, and now
well known ones, Point of Rocks and Green River. In a table of distribution the data are assumed to exist to justify this classification.

Notwithstanding these efforts to sustain the argument for the Tertiary age of the central coal formation of the West, it had been so weakened by the blows of King and Marsh, coupled with the admissions of
· Meek, that little remained but the evidence afforded by the fossil plants
in its support, and this, though abundant in quantity, was naturally dis-
trusted, and had been enfeebled by the considerations urged against it
by Le Conte. Meek himself did not hesitate to refer forms of Ostrea
and Anomia, from Point of Rocks on the Union Pacific Railroad and in
the typical Bitter Creek district, to the Cretaceous,[1] and now there was
destined to come forward a new discovery of great importance, the full
weight of which fell upon that side of the question. In the summer of
1872 Messrs. Meek and Bannister discovered the bones of a large saurian
near Black Buttes Station in the Bitter Creek series, and Professor Cope
soon after visited the spot and studied the fossils. He laid his results
before the American Association for the Advancement of Science at
Dubuque in August of that year, and published his descriptions in the
Proceedings of the American Philosophical Society for September 19.
In this paper he remarks (p. 483) : " From the above description it is evident that the animal of Black Buttes is a Dinosaurian reptile. • • •
It is thus conclusively proven that the coal strata of the Bitter Creek
Basin of Wyoming Territory, which embraces the greater area yet discovered, were deposited during the Cretaceous period, and not during
the Tertiary, though not long preceding the latter." And, commenting
upon the same subject in the American Naturalist for November, 1872, he
says : " This discovery places this group without doubt within the limits
of the Cretaceous period."

Mr. Lesquereux was also in the field this year (1872), and his investigations, at the request of Dr. Hayden, were specially directed to " posi-

[1] Fifth Annual Report United States Geological Survey of the Territories, 1871, p.
375.

tively ascertaining the age of the lignitic formations." He visited most of the important points in Wyoming, Colorado, and New Mexico, and prepared an elaborate report, in which, it is needless to say, he confirmed and reasserted his former conclusions as to the Tertiary age of the entire coal-bearing series, which he denominates the American Eocene.[1]

The reports of Messrs. Meek and Bannister were also published in the same volume. The former expresses himself with his usual caution, admitting that the invertebrate fossils were inadequate to determine the age of this group, and that his former reference of certain species to the Cretaceous was not prompted by the evidence afforded by the forms themselves (pp. 457, 458). Some of the statements made in this report have acquired special interest in the light of recent investigations and in view of the gradual settlement of opinion which seems to be now going on respecting this much discussed question. He says (p. 460): "The most surprising fact to me, supposing this to be a Cretaceous formation, is, that we found directly associated with the reptilian remains at Black Buttes a shell I cannot distinguish from *Viviparus trochiformis*, originally described from the Lignitic formation at Fort Clarke, on the Upper Missouri, a formation that has always been regarded as Tertiary by all who have studied its fossils, both animal and vegetable. * * * The occurrence of this last mentioned species here, along with a Cretaceous type of reptilian, and a *Corbicula* apparently identical with *C. cytheriformis* of the Judith River brackish-water beds, together with the presence of Corbiculas very closely allied to Judith River species, at lower horizons in this series, and the occurrence of some vertebrates of Cretaceous affinities at the Judith River localities, would certainly strongly favor the conclusion, not only that this Judith formation, the age of which has so long been in doubt, is also Cretaceous, but that even the higher fresh-water lignite formation at Fort Clarke and other Upper Missouri localities may also be Upper Cretaceous instead of Lower Tertiary."

From these and other expressions in this report Mr. Meek may be fairly said to have conceded the Cretaceous age of the Bitter Creek series, but he insists that the Judith River deposits must go with it into that formation, while of the Fort Union group his position may be summed up by quoting his remark that it would take very strong evidence to convince him "that the higher fresh-water Lignite series of the Upper Missouri is more ancient than the Lower Eocene."

The year 1874 found the discussion of the age of the so-called American Lignitic at its height. A paper in the American Journal of Science for April of that year, by Dr. Newberry, and a reply to it by Mr. Lesquereux in the same journal for June, deserve special attention. The former makes bold to say that to his "certain knowledge" a considerable portion [that of New Mexico] of the flora which the latter had called

---

[1] Annual Report of the United States Geological Survey of the Territories, 1872, pp. 339, 343.

Eocene in his last report is Cretaceous, and that another considerable
portion [that of the Upper Missouri] is of Miocene age, and he denies
that the flora of any part of the American coal series possesses an Eocene facies. Mr. Lesquereux's reply is of course a defense of his former
position and is supported by a vast array of facts.

In the first bulletin of the Geological Survey of the Territories, published in 1874, Professor Cope, from evidence supplied by vertebrate
remains, refers the Great Lignitic of the Upper Missouri to the same
section of geologic time as the Bitter Creek coal series, now settled in
his mind as Cretaceous, and in Bulletin No. 2 (pp. 5–19) appeared an
elaborate report by the same author (reproduced, apparently without
change, in the Annual Report for 1873, also published in 1874 and
later than the Bulletins, pp. 431–446), in which he sums up the evidence
from the side of vertebrate paleontology. In this report Professor Cope
gives Mr. Lesquereux full credit for accurately co-ordinating the data
furnished by the vegetable remains, and concludes "*that a Tertiary
flora was contemporaneous with a Cretaceous fauna, establishing an unin-
terrupted succession of life* across what is generally regarded as one of
the greatest breaks in geologic time." His further remark that "the
appearance of mammalia and sudden disappearance of the large Mesozoic types of reptiles may be regarded as *evidence of migration and not
of creation*," embodies a thought that has been since revived and extended.

To this report of Professor Cope, as published in the Annual Report
for 1873, he appends a short discussion, not contained in the Bulletin,
in the nature of a reply to the article of Dr. Newberry above referred
to. In the course of this discussion the following remarks occur: " If
a flora below the Cretaceous of New Mexico resembles a Tertiary one,
how much more probable is it that the floras of the Lignites of Colorado
and Wyoming are such, as they are known to be of later age than those
of New Mexico, and to be at the summit of the Cretaceous series, as indicated by animal remains; and if the flora of the Fort Union beds be
Miocene, that of the identical horizon in Colorado must be Miocene
also; and if the vegetation below this flora be so distinct from it, what
is more probable, according to the evidence adduced by Dr. Newberry,
than that they are Eocene, as maintained by Mr. Lesquereux? That
such should be the case is in harmony rather than in conflict with the
facts presented by the existing life of the earth, where we have the
modern fauna of the northern hemisphere contemporary with a partly
Eocene and partly Mesozoic fauna in the southern."

The same volume contains a report by Mr. Archibald Marvine of his
operations during the season of 1873 in the park districts of Colorado.
In treating the " Lignitic formation," as observed by him, he reviews
the evidence from the plant remains, as interpreted by Lesquereux, as
well as that furnished by vertebrate life, and says: " It must be supposed, then, that either a Cretaceous fauna extended forward into the

Eocene period, and existed contemporaneously with an Eocene flora, or else that a flora wonderfully prophetic of Eocene times anticipated its age and flourished in the Cretaceous period to the exclusion of all Cretaceous plant forms. * * * In either case, the fact remains that here the physical and other conditions were such that one of the great kingdoms of life, in its progress of development, either lost or gained upon the other, thus destroying relations and associations which existed between them in those regions from which were derived the first ideas of the life boundaries of geological time, causing here apparent anomalies." He adds the following important paragraph: "Much of the confusion and discrepancy has, in my opinion, arisen from regarding different horizons as one and the same thing. It must be distinctly understood that this group as it exists east of the mountains in Colorado is very different from, and must not be confounded with, the horizon in which much of the Utah and New Mexican lignite occurs, and which belongs undoubtedly to the Lower Cretaceous; and, further, that the extended explorations of Hayden and others would seem to prove almost conclusively that the Colorado lignitic group is the direct southern stratigraphical equivalent of the Fort Union group of the Upper Missouri, which is considered generally to be no older than the Eocene, while Newberry asserts it to be Miocene."

Mr. Lesquereux returns again, in his contribution to this same volume, to the defense of his former position. He disposes in a manner of the statement that characteristic Cretaceous molluscan fossils had been found "above the beds of the lignitic formations" by quoting Messrs. Cox and Berthoud, the collectors of the specimens about which so much had been said, who both show that the conditions under which they occurred were such as to render their stratigraphical position too doubtful to form the basis for such important generalizations. He reasserts his belief in "the unity of the Lignitic formation in its whole," and reargues the whole case. He also revises his "groups" and gives lists of all the species found in each.

In Volume VII of the Canadian Naturalist, p. 241, published in 1874, Mr. George M. Dawson discusses "The Lignite Formations of the West," now discovered to extend far up into Canadian territory. He regards them as of later age than the Cretaceous and accepts the view of Messrs. Hayden and Lesquereux that the Fort Union group is Eocene. Referring to the opinions of Cope, he says: "The evidence does not appear to show that the Cretaceous species were of themselves becoming rapidly extinct, but that over the Western region, now forming part of this continent, the physical conditions changing drove the Cretaceous marine animals to other regions, and it is impossible at present to tell how long they may have endured in oceanic areas in other parts of the world. This being so, and in view of the evidence of the preponderant animal and vegetable forms, it seems reasonable to take the well marked base of the Lignite series as that of the lowest Tertiary, at least at present. The formation described belongs to this lowest Tertiary, being, in fact, an extension of Hayden's *Fort Union group*, and from analogy may be called *Eocene*."

In a more formal paper[1] published the same year, he also says: "The formation is, however, undoubtedly an extension of the Great Lignite or Fort Union group of strata of Hayden, as developed in the Western States and Territories. * * * These strata immediately succeeding the Cretaceous rocks are the lowest American representatives of the Tertiary series and have been called for this reason Eocene, though it is impossible to affirm that their deposit was more than approximately synchronous with that of the Eocene as constituted in Europe" (p. 20).

Returning to the same subject a year later in his final report of the Northwest Boundary Commission,[2] after familiarizing himself with the discussions going on in the United States, the same author adheres to his previous views and remarks: "There seems little doubt, however, that the general tenor of the evidence of these beds, when considered alone, favors their Lower Eocene age. Their exact synchronism with the European Eocene is a question apart from the present inquiry" (p. 186).

Early in 1875 Professor Cope, who had examined the vertebrate remains sent him by Mr. Dawson from near Milk River, on the boundary of the British possessions, published a note upon them,[3] in which he says: "The genus of tortoises *Compsemys*, Leidy, is peculiar to the Fort Union epoch, while *Plastomenus*, Cope, belongs to the Eocene. Its presence in this fauna would constitute an important assimilation to the Lower Tertiary, but the specimens are not complete in some points necessary to a final reference. The species are in any case nearly allied to that genus. There are, however, gar scales included in the collection which closely resemble those of the genus *Clastes* of the lower Eocenes of the Rocky Mountains. This is empirically another indication of near connection with Tertiary time, but not conclusive, since allied genera have a much earlier origin in Mesozoic time. * * * Nevertheless, the list of species, short as it is, indicates the future discovery of a complete transition from Cretaceous to Eocene life more clearly than any collection yet obtained marking this horizon in the West."

[1] Report on the Tertiary Lignite Formation in the Vicinity of the forty-ninth parallel. By George M. Dawson. Addressed to Capt. D. R. Cameron, R. A., H. M. Boundary Commissioner. British North American Boundary Commission. Geological Report of Progress for the year 1873 [in part]. Montreal, 1874.

[2] British North American Boundary Commission. Report on the Geology and Resources of the Region in the Vicinity of the forty-ninth parallel, from the Lake of the Woods to the Rocky Mountains, with lists of plants and animals collected, and notes on the fossils. By George Mercer Dawson, geologist and botanist to the Commission. Addressed to Maj. D. R. Cameron, R. A., H. M. Boundary Commissioner. Montreal, 1875.

[3] Proceedings of the Academy of Natural Sciences, Philadelphia, Jan. 5, 1875, Vol. XXVII, pp. 9, 10.

Professor Cope's article, from which we made quotations a few pages back, appeared for the third time in his final report on fossil vertebrates[1] with very few changes. It is to be noted, however, that he no longer proposes to call the lignite deposit the sixth member of the Cretaceous formation of the West, and referring to the fossils from the Milk River district last mentioned we find him saying "that there are present two genera in this collection which are diagnostic of the Fort Union epoch, but no species certainly so, though two species are probably identical with species of that epoch; also * * * that the species referred to *Plastomenus* constitute an indication of affinity with corresponding Eocene forms. The presence of gar fishes of the genus *Clastes* in this formation is as yet peculiar to this and the Judith River localities. As these gars have not heretofore been found in North America below the Eocene, they constitute the first case of apparent commingling of Tertiary and Cretaceous animal life yet clearly determined." He is careful to add, however, that the evidence of the Dinosaurs outweighs these considerations.

At this time, when at least one vertebrate paleontologist was beginning to concede that this formation, though apparently Mesozoic, yet possessed a marked Tertiary facies, Mr. John J. Stevenson came forward with several papers[2] from the stratigraphical side in support of the Cretaceous theory. His language is the most positive of any yet employed, but a careful examination of his statements shows that his argument acquired its chief force from the form in which it was put forward. Such statements as that "everywhere the sandstones of the Upper Cretaceous present the same lithological character;" that "not a single Tertiary species occurs in the whole series;" that "wherever animal remains occur with this fucoid [Halymenites] they are invariably characteristic Cretaceous species;" that "the evidence in favor of Cretaceous age is abundant;" that the record of plant life is "little better than a blank, with here and there a few markings, many of which are too indistinct to be deciphered;" that "the only fossils characteristic of No. 5 ever obtained from Colorado were procured from rocks which are most probably the very highest strata of the Lignitic series"—would, if the question were at all one of credibility, as it is not, clearly invalidate this witness and make his own charge, "*falsus in uno, falsus in omnibus*," peculiarly applicable to himself. Mr. Stevenson's writings, however, have the merit of defending the essential unity of all the lignitic deposits.

---

[1] Report of the United States Geological Survey of the Territories, Vol. 11, 4-, 1875, pp. 25–41.

[2] Proceedings of the Society of Natural History, New York, 2d ser., No. 4, 1874, p. 93; Age of the Colorado Lignites, Reports upon Geographical and Geological Exploration and Survey West of the One Hundredth Meridian, in charge of First Lieut. Geo. M. Wheeler, Vol. III, 1875, pp. 404–410; The Geological Relations of the Lignitic Groups, Proceedings of the American Philosophical Society, Vol. XIV, pp. 447–475.

The Annual Report of the Geological Survey of the Territories under Dr. Hayden for 1874, published in 1876, contains three very important papers upon this subject. The first is by Dr. Hayden himself, who labors effectively to "connect the coal-bearing beds of the Laramie Plains and Colorado with the vast group in the Northwest," but concedes the Cretaceous age of the Bear River and Coalville deposits. He says that "above the upper Fox Hills group there are about 200 feet of barren beds which may be regarded as beds of passage to the Lignitic group, which more properly belong with the Fox Hills group below. In this group of transition beds all trace of the abundant invertebrate life of the great Cretaceous series below has disappeared. * * * Whatever view we may take with regard to the age of the Lignitic group, we may certainly claim that it forms one of the time boundaries in the geological history of our western continent. It may matter little whether we call it Upper Cretaceous or Lower Eocene, so far as the final result is concerned. * * * Even the vertebrate paleontologists, who pronounce with great positiveness the Cretaceous age of the Lignitic group, do not claim that a single species of vertebrate animal passes above the horizon I have defined from the well marked Cretaceous group below."

The second of these papers is by Dr. A. C. Peale, who has here performed good service in preparing tables to illustrate the progress of opinion on this subject. In addition to this, however, after stating the character of his own investigations, he gives it as his opinion that "the lignite-bearing beds east of the mountains in Colorado are the equivalent of the Fort Union group of the Upper Missouri, and are Eocene-Tertiary; also, that the lower part of the group, at least at the locality two hundred miles east of the mountains, is the equivalent of a part of the lignitic strata of Wyoming;" but he thinks that "the Judith River beds have their equivalent along the eastern edge of the mountains, below the Lignite or Fort Union group, and also in Wyoming, and are Cretaceous, although of a higher horizon than the coal-bearing strata of Coalville and Bear River, Utah. They form either the upper part of the Fox Hills group (No. 5) or a group to be called No. 6."

Finally we have another exhaustive paper by Mr. Lesquereux, in which he divides the arguments against the Tertiary theory into five propositions and answers each in detail. Important discoveries of fossil plants had been made during the year at Point of Rocks, and these are made to lend their weight to his argument. It is needless to say that his conclusions remained unchanged.

The ninth volume of the final quarto reports of the Geological Survey of the Territories, consisting of Mr. Meek's report on the invertebrate Cretaceous and Tertiary fossils of the Upper Missouri country, appeared in 1876. In this report Mr. Meek takes the ground that the Judith River beds are distinct from the Fort Union group proper and of Cretaceous age, or at least probably so; but he is inclined to believe, from the occurrence of similar forms in both, that they are the equivalent of the Bitter Creek series in Wyoming. As to the Fort Union beds, he adheres to his former opinion, that they represent the lower Eocene. He deprecates the attempt to unify all the lignite-bearing rocks, and remarks: "The presence or absence of lignite proves nothing of itself, as lignite undoubtedly occurs in both Cretaceous and Tertiary rocks in the far West." In his comparisons of the Fort Union with the Wyoming deposits he states that the species of the former are all different from those of the Bitter Creek group, and concludes that these groups at least cannot be equivalents. Mr. Meek's concluding remarks upon the conflicting testimony of fossils and its lessons (pp. lx, lxi) are a model of scientific reasoning, and doubtless went far to mitigate the acerbity of this prolonged debate.

Powell's Geology of the Uintah Mountains was published the same year (1876) as the report last mentioned, and contains an important contribution to the present subject. Professor Powell and Dr. C. A. White had gone carefully over the disputed ground of the Bitter Creek district, tracing it up to its junction with the Washakie and Green River beds on the west, and in this volume both these authorities record their conclusions, which are in substantial accord. The former remarks (p. 67): "The relation of these groups to those established by Professors Meek and Hayden on the Upper Missouri is not well determined. * * * All the evidence that has been published by Dr. Hayden and members of his corps concerning the Park Province, and all my own observations in that region, lead me to the conclusion that a long chain of islands stretched in a northerly and southerly direction through that region of country, separating the Cretaceous sea of the Plateau Province from the Cretaceous sea of the Upper Missouri."

Between Black Buttes Station and Point of Rocks Station, on the Union Pacific Railroad, these gentlemen discovered a "physical break" in the series, exposing at the latter point a lower formation; and at this point they fixed the line between Mesozoic and Cenozoic strata, assigning, in the table of groups on page 40, the Point of Rocks group to the Cretaceous and the Bitter Creek group to the Tertiary. On this subject Professor Powell says (p. 71): "On account of the discussions which have arisen concerning the age of certain beds of lignitic coal, the plane of demarkation between the Cenozoic and Mesozoic may subject me to criticism; but, geologically, the plane is important, as it represents a decided physical change, and it certainly harmonizes with the opinion of paleontologists to a degree that is somewhat surprising. All of the plants described by Professor Lesquereux and collected by himself and others within this province have been referred by him to divisions in the Tertiary, and are found in strata above this physical break, and hence I agree with him in considering them Tertiary. * * * The conclusions reached from a study of the vertebrate paleontology by Professors Leidy, Marsh, and Cope entirely harmonize with this division of the Cenozoic and Mesozoic. There is a single exception to this: Professor

Cope described a *Dinosaur* found near Black Buttes Station as Cretaceous. I have verified the determination of the stratigraphic horizon
by examining the place and finding other *Dinosaur* bones; but this horizon is above the physical break, and the evidence of the *Dinosaur* seems
to be contradicted by the evidence furnished by many other species
described by Professor Cope from about the same horizon."

Dr. White also discusses this question in the same volume, and states
his reasons for regarding the Point of Rocks beds as Cretaceous in the
following words (pp. 83, 84): "There is no physical break between this
group and the Salt Wells group below it. Its strata contain at least
three species of Inoceramus, which genus has never been known in
strata of later date than the Cretaceous period. Odontobasis, a species
of which has been obtained from near the summit of the group, is regarded as a Cretaceous genus; and in view of the facts before stated,
that land and fresh- and brackish-water mollusks are comparatively
valueless as indices of the passage of geological time, the presence of
no known forms in its strata forbids the reference of this group to the
Cretaceous period."

On the other hand, the Bitter Creek series proper is referred to the
Eocene, and to the question "Why has the dividing line between the
strata of the Tertiary and Cretaceous periods been drawn where it is
rather than at some horizon either above or below it?" his answer is:
"There is no physical break in the Cretaceous strata from the base of
the series to the top of the upper, or Point of Rocks group, at which
horizon there is at all observed points, extending over a large region, a
considerable unconformability by erosion of the lower strata of the
Bitter Creek group upon the upper strata of the Point of Rocks group
(p. 87)."

The second volume of the Reports of the Geological Exploration of the
Fortieth Parallel by Mr. Clarence King, which appeared in 1877, contains
exhaustive papers upon the geology of this region by Messrs. Arnold
Hague and S. F. Emmons, who had studied the rocks with great care.
Both these gentlemen agree in referring the entire lignite-bearing series
to the Cretaceous. They do not draw the nice distinction made by
Messrs. King, Powell, and White, but Mr. Hague seems to have no
doubt that even the Carbon coals belong there, while Mr. Emmons similarly disposes of those of Evanston. In this report the term *Lignitic*
is abandoned altogether and the term *Laramie* is applied to this formation. Mr. Emmons constantly speaks of the "Laramie Cretaceous" and
the "Laramie group," the latter of which terms has now been generally
adopted and extended over a much wider area.

In his vice-presidential address, delivered before the American Association for the Advancement of Science, at Nashville, Tenn., August 30,
1877, Prof. O. C. Marsh expressed himself as follows upon the general
subject under discussion: "The boundary line between the Cretaceous
and Tertiary in the region of the Rocky Mountains has been much in

dispute during the last few years, mainly in consequence of the uncertain geological bearings of the fossil plants found near this horizon. The accompanying invertebrate fossils have thrown little light on the question, which is essentially, whether the great lignite series of the West is uppermost Cretaceous or lowest Eocene. The evidence of the numerous vertebrate remains is, in my judgment, decisive, and in favor of the former view."[1]

At about this time the researches of Dr. C. A. White, who had become deeply interested in this formation, began to bring forth important results. His "Paleontological Papers" commenced to appear in 1877, as contributions to the Bulletins of Dr. Hayden's Survey, in the third of which he drew up tables of the groups of the Green River and Upper Missouri River regions. It was here that he employed the term "Post-Cretaceous," to include the Laramie group of the King Reports and the lower third of the Wasatch group, and correlating the Judith River with the Laramie and the Fort Union with the Wasatch group. In the fifth of these papers, published the same year, he enters more fully into the discussion of the age of these groups and remarks: "With a few doubtful exceptions, none of the strata of the Laramie group were deposited in open sea waters; and, with equally few exceptions, none have yet furnished invertebrate fossils that indicate the Cretaceous rather than the Tertiary age of the group. These latter exceptions are some *Inocerami* that have been obtained upon the lower confines of the group, and doubtfully referred to it rather than to the Fox Hills group below; and also a species of *Odontobasis* from strata near the top of the group, two miles west of Point of Rocks Station, Wyoming. The latter genus, established by Mr. Meek, is comparatively little known, but it was regarded by him as characteristic of the Cretaceous period. This constitutes the slender evidence of the Cretaceous age of the Laramie group that invertebrate paleontology has yet afforded.

"Again, the brackish- and fresh-water types of *Mollusca* that are afforded by the Laramie and the lower portion of the Wasatch group are in most cases remarkably similar, and some of the species of each group respectively approach each other so nearly in their characteristics that it is often difficult to say in what respect they materially differ. Moreover, they give the same uncertain indication as to their geological age that all fossils of fresh- and brackish-water origin are known to do.

"It is in view of the facts here stated, and also because I believe that a proper interpretation of them shows the strata of the Laramie group and the base of the Wasatch to be of later date than any others that have hitherto been referred to the Cretaceous period, and also earlier

[1] Proceedings of the American Association for the Advancement of Science, 1877, page 229.

than the Eocene epoch, that I have decided to designate those strata as Post-Cretaceous, at least provisionally."

By a remarkable coincidence this term *Post-Cretaceous* was applied to the lignitic beds of the Trinidad district, New Mexico, by Dr. F. M. Endlich, in the Annual Report of the Geological Survey of the Territories for 1875 (p. 206), published in 1877; but it is impossible to say which of these reports should have priority, and as the term has now been generally abandoned this is quite unimportant.

In the death of Mr. F. B. Meek the science of invertebrate paleontology lost one of its ablest votaries, and but for the fact that Dr. C. A. White had already entered the field in this rôle as well as in that of stratigraphical geologist, this department of research in our western formations might have been sadly neglected. But the now rapidly increasing writings of the latter author fully supplied the place of the former, and the contest went on. In the Annual Report of Dr. Hayden's Survey for 1876, published in 1878, Dr. White reports his operations during the years 1876 and 1877 in Colorado, in which paper he takes occasion to draw up a section of the rocks and to prepare a table of correlated general sections which are highly instructive. Confining ourselves to the Laramie group, we see that he adopts that term and makes it commensurate with his Post-Cretaceous, to which he still adheres, and also with the Laramie of King and the Lignitic of Meek and Hayden. The Point of Rocks group of Powell begins with the Laramie, but stops at a lower horizon, his Bitter Creek group occupying the remainder, and the whole of the Wasatch (the Vermilion Creek group of King). In defense of his course in receding from his former position, in which his views agreed with those of Powell, he says: "After a careful examination of the extensive exposures of this series of strata, as well as those of the Wasatch group above it in this district, I have failed to discover any unconformity such as exists in the valley of Bitter Creek. Therefore, the greatest unconformity that is now known to exist among any of the strata from the base of the Cretaceous to the top of what I here designate as the Post-Cretaceous, is found among the strata of the latter group, and not at its top. In this district and the region immediately adjoining it, whatever catastrophal or secular changes may have meanwhile taken place elsewhere, or even extending within its limits, sedimentation was evidently continuous and unbroken, not only through this series itself, but also into and through the whole Wasatch group.

"The fact that this series passes insensibly into the Fox Hills group below, and into the Wasatch group above, renders it difficult to fix upon a stratigraphical plane of demarkation, either for its base or summit. I have, therefore, decided to regard this group as essentially a brackish-water one, referring all strata below that contain any marine Cretaceous invertebrate forms to the Fox Hills group, beginning this series with those strata that contain brackish- and fresh-water forms, and ending it above with those strata in which the brackish-water forms finally cease. Thus defined, the whole series seems to form one natural paleontological group, as well as to be a sufficiently distinct stratigraphical one, for which I have adopted the name of Laramie group of King."

In giving his reasons for adhering to the name Post-Cretaceous, Dr. White further says: "The flora of this group is understood to be wholly of Tertiary types, according to Professor Lesquereux. None of its invertebrate fossils are of distinctive Cretaceous types, although fossils of similar types are known to occur in Cretaceous as well as Tertiary strata. So far, then, as the flora and invertebrate fauna are concerned, there is nothing to indicate the Cretaceous age of the group. In fact, invertebrate paleontology is utterly silent upon the subject. On the contrary, Professor Cope finds reptilian remains, even in the uppermost strata of the group, that he regards as of Cretaceous type. I believe that, upon the evidence of invertebrate paleontology, the Fox Hills group is later than the latest Cretaceous strata of Europe; and I therefore regard the Laramie group as occupying transitional ground between the well marked Cretaceous and Tertiary groups, but this opinion is only tentatively held until further facts are obtained."

The term Post-Cretaceous is employed by both Endlich and Peale in their reports in this volume (pp. 77, 109, 181).

In his seventh Paleontological Paper (Bulletin U. S. Geological Survey of the Territories, Vol. IV, No. 3), distributed in 1878, Dr. White greatly extends the boundaries of the Laramie group, making it embrace "both the Judith River and Fort Union series of the Upper Missouri River; the Lignitic series east of the Rocky Mountains in Colorado; the Bitter Creek series of Southern Wyoming and the adjacent parts of Colorado; and also the 'Bear River estuary beds,' together with the Evanston coal series of the valley of Bear River and adjacent parts of Utah," as well as strata known to exist in other large and widely separated districts of the western portion of the National domain, and he gives a list of species characteristic of the group, showing their distribution throughout these several districts.

Mr. Leo Lesquereux's so-called "Tertiary Flora" constitutes the seventh volume of the final reports of the Geological Survey of the Territories under Dr. F. V. Hayden, which, of course, embraces the plants of the Laramie group. In it Mr. Lesquereux has given full scope to the expression of his views upon the age of this group, and it is naturally here that we must look for the most able and exhaustive treatment of the subject thus far presented by this author. In the letter of Dr. Hayden to the Secretary of the Interior transmitting this report, and which bears date January 1, 1878, he again reviews this subject and remarks: "The author states that his final conclusions do not differ materially from those already advanced by myself, and he regards the evidence as conclusive that the Lignitic group is of Tertiary age. This result is gratifying, not only as settling the question at issue, but as silencing criticism of the value and reliability of the general work accomplished by the survey under my direction." But in this same letter Dr. Hayden also declares his conviction, more than once before expressed, but not as yet, so far as I know, accepted by either Lesquereux or Newberry, " that the Fort Union beds of the Upper Missouri River are the equivalent of the Lignitic formation as it exists along the base of the Rocky Mountains, in Colorado," as well as of the Bitter Creek series west of the Rocky Mountains, as argued by Dr. White, and he says: " It is also probable that the brackish-water beds on the Upper Missouri must be correlated with the Laramie, and that the Wahsatch group as now defined and the Fort Union group are identical as a whole, or in part at least."

As Mr. Lesquereux's conclusions expressed in this report are the same as he had held throughout the discussion, and the arguments not new, no further elucidation of them is necessary.

Volume I of Mr. Clarence King's Geological Reports of the Survey of the Fortieth Parallel, treating of the systematic geology, and written by Mr. King himself, did not appear until 1878. His views upon this question were looked for with great interest, though it was, of course, to be expected that they would coincide generally with those of his assistants already published in other volumes. Notwithstanding the tendency, which had been marked for several years, to regard the attempt to assign the Laramie group to either the Cretaceous or Tertiary age as not only profitless but rather puerile, inasmuch as its relative position in the western American system was so well settled, Mr. King did not consider it beneath the dignity of this stately report to approach the subject much from the old standpoint and record his position in nearly conventional terms. He says (p. 350): "Aside from the Taconic system, no single geological feature in all America has ever given rise to a more extended controversy than the true assignment of the age of this group. On data which will presently be set forth, it is assumed by us to be the closing member of the Cretaceous series, and the last group of the great conformable system which east of the Wahsatch stretches upward from the base of the Cambrian."

The views that had been put forth in opposition to this he then arranges into a series of seven "assumptions," which he proceeds to consider and dispose of in the order laid down. As some of these points are admitted and others not vital, they need not be noticed *seriatim* ; a few extracts must suffice. He says (p. 352): "A complete refutation of assumption three, that the fauna proves a Tertiary, not a Cretaceous age, is found in the fact that the evidence of a meagre molluscan life and a large range of plants cannot be held to weigh against the actual presence of *Dinosauria* in the very uppermost Laramie beds, and, as will appear in the sequel, of an abundant *lowest Eocene* mammalian fauna in the unconformably overlying Vermilion Creek group. * * * As-

sumption number five, as to the conformity of the Laramie with the
Wahsatch group, I shall presently proceed to show, is based upon imperfect knowledge, and is abundantly disproved by repeated sections."

Relative to the Fort Union group, he admits that he had never visited
that locality, but notes the conflicting evidence of vertebrate and vegetable remains, and Mr. Lesquereux's silence upon the latter in his Tertiary
Flora, and remarks (p. 353) that " the further correlation of the upper
plant-beds of Fort Union with the Wahsatch (my Vermilion Creek) seems
the most prodigious strain. The Wahsatch (Vermilion Creek), or un-
mistakable lowest Eocene, is nonconformable with the Laramie. The
relations of conformity or nonconformity between the plant-bearing
beds of Fort Union and the Dinosaurian beds are not given, and there
is reason to believe that the plant beds represent a horizon of the great
White River Miocene series, which underlies the Pliocene over so large a
part of the Great Plains. * * * I apprehend that the plant horizon
at Fort Union will be found to be nothing but the northward extension
of the White River Miocene."

Professor Cope's paper on horizons of extinct vertebrata, in the fifth
volume of the Bulletins of the United States Geological and Geographical Survey of the Territories (No. 1, Art. 11), which appeared early
in the year 1879, is of special value as the first attempt to correlate the
Laramie group with European strata upon the evidence of vertebrate
remains. This discussion was repeated without essential change in his
great work which forms Book I of the third volume of the final quarto
reports of that Survey, published in 1884. The general result is a still
further yielding on the part of the writer to the views of the invertebrate and vegetable paleontologists against the decidedly Cretaceous
character of the group. He shows in an instructive way that it bears
a very close relation to the Sables of Brachenx and Conglomerates
of Cerny, which are Eocene, but with this difference, " that the characteristic genera of reptiles and fishes of the Laramie of North America are in America associated with Cretaceous *Dinosauria* and not
with *Mammalia ;* while in Europe they are associated with *Mammalia* and
not with *Dinosauria*." And he adds: " In arranging the Laramie group,
its necessary position is between Tertiary and Cretaceous, but on the
Cretaceous side of the boundary, if we retain those grand divisions, which
it appears to me to be desirable to do;" and he admits " that another
formation must be added to the series already recognized in France,
viz, the Laramie, or Post-Cretaceous." This he does in his table of
correlated general sections, on page 50, making the Post-Cretaceous
embrace the Laramie and the Puerco, the former in turn being equivalent to the combined strata of the Judith River and Fort Union
deposits.

Dr. C. A. White's elaborate report upon his extensive field researches
made in 1877 appeared in the Annual Report of the Geological Survey of
the Territories for that year, which, however, did not see the light till

1879. Dr. White had spent the entire season in the exhaustive study of the various outcrops of the Laramie in Colorado and Wyoming on both sides of the Rocky Mountains, and had made large and valuable collections, which he had worked up with care, and which form the substantial basis for his conclusions as here set forth. In his "general discussion," which follows the detailed report, starting with "the unity of all the principal brackish-water deposits hitherto known in the Western Territories, and * * * their recognition as a comprehensive group of strata under the name of the Laramie group, which represents a great period in geological time, and especially such in the geological history of North America," he proceeds to discuss, not so much the *age* of the group, as the conditions of its deposition and the geological history of the western part of the continent following the close of true Cretaceous time. Into this discussion, though confessing its superior importance, we cannot here enter, but must be content to cite a passage or two to show to what conclusions he had now come relative to the age of the Laramie group, its geographical boundaries, and the thickness of its deposits. He says:

"Resting directly upon the strata of the Fox Hills group are those of the Laramie group, the latter, as already shown, having been, at least in part, deposited continuously with the former. The geographical boundaries of the great Laramie formation are not known, but its area embraces many thousand square miles, for it is known to extend from Southern Colorado and Utah northward beyond the northern boundary of the United States, and from the Wahsatch Mountains eastward far out on the great plains. It reaches a maximum thickness of about 4,000 feet, and its general lithological characteristics are similar to those of the Fox Hills group, a known marine formation. Its fauna, however, has been shown to be largely of brackish- and partly of fresh-water origin, and not marine. Furthermore, the brackish-water species are distributed throughout its entire thickness and its whole geographical extent. These facts, together with the absence from all the strata yet examined of any true estuary characters, show that the Laramie group was deposited in a great brackish-water sea. * * *

"In the foregoing report I have purposely avoided an expression of opinion as to the true geological age of the Laramie group, because, notwithstanding the positive opinions that have been expressed by others upon that subject, I regard it as still an open question. * * * The claim that Cretaceous types of vertebrates are found in even the higher strata of the Laramie group is freely conceded, and I have no occasion to question the reference that has been made of its fossil plants, even those of the lowest strata, to Tertiary types. The invertebrate fossils of the group itself, as I have elsewhere shown, are silent upon this subject, because the types are either unique, are known to exist in both Mesozoic and Tertiary strata, or pertain to living as well as fossil forms. Every species found in the Laramie group is no doubt extinct, but

the types have collectively an aspect so modern, that one almost in-
stinctively regards them as Tertiary; and yet some of these types are
now known to have existed in the Cretaceous and even in the Jurassic
period.

"In view of the conflicting and silent character, respectively, of these
paleontological oracles the following suggestions are offered : It is a
well-known fact that we have in North America no strata which are,
according to European standards, equivalent with the Lower Cretaceous of Europe, but that all North American strata of the Cretaceous
period are equivalent with those of the Upper Cretaceous of that part
of the world. That the Fox Hills group is of Upper Cretaceous age
no one disputes, the only question being as to its place in the series.
A comparison of its fossil invertebrate types with those of the European Cretaceous indicates that it is at least as late as, if not later than,
the latest known Cretaceous strata in Europe. If, therefore, that par-
allelism is correctly drawn, and the Laramie group is of Cretaceous
age, we have represented in America a great and important period of
that age which is yet unknown in any other part of the world. Besides this, we may reasonably conclude that the Fox Hills group of
the West is equivalent with the Upper Cretaceous strata of the Atlantic and Gulf coasts, between which and the Eocene Tertiary of those
regions there is no known equivalent of the Laramie group.

"If paleontologists should finally agree upon regarding the Laramie
group as of Cretaceous age, it must be because of the continuance of
certain vertebrate Cretaceous types to the close of that period, and
the presence of mammalian Tertiary types in the strata immediately
following; but the following facts, in addition to those which have been
already stated, should be carefully considered before any such agreement is made :

"With rare and obscure exceptions no mammalian remains are known
in North American strata of earlier date than that of those which were
deposited immediately after the close of the Laramie period and upon its
strata. Immediately from and after the close of the Laramie period
their abundant remains in the fresh-water Tertiaries of the West show
that highly-organized mammals existed in great variety and abundance; all of which may be properly regarded as constituents of a Tertiary fauna, and many of which are by accepted standards of distinctively Tertiary types. If the presence of these forms in the strata referred to, and their absence from the Laramie strata immediately beneath them, together with the presence of Dinosaurians there, be held
to prove the Tertiary age of the former strata, then was the Tertiary
period ushered in with most unnatural suddenness. Sedimentation was,
at least in part, unbroken between the Laramie group and the strata
which contain the mammalian remains referred to, so that the local conditions of the origin of all of them were substantially the same, and

yet, so far as any accumulated evidence shows, those mammalia were not preceded in the Laramie period by any related forms. Such suddenness of introduction makes it almost certain that it was caused by the removal of some physical barrier, so that ground which was before potentially Tertiary became so by actual faunal occupancy. In other words, it seems certain that those Tertiary mammalian types were evolved in some other region before the close of the Laramie period, where they existed contemporaneously with at least the later Laramie Dinosaurians of Cretaceous types, and that the barrier which separated the faunæ was removed by some one of the various movements connected with the evolution of the continent. The climate and other physical conditions which were essential to the existence of the Dinosaurians of the Laramie period having evidently been continued into the Tertiary epochs that are represented by the Wahsatch, Green River, and Bridger groups, they might doubtless have continued their existence through those epochs as well as through the Laramie period, but for the irruption of the mammalian horde, to which they probably soon succumbed in an unequal struggle for existence."

From the above extracts it will at once be seen that Dr. White had now succeeded in raising this discussion from the comparatively trivial question as to the name which should be given to the age occupied by the Laramie group to one involving not only the manner in which the continent was formed, but also the origin, development, extinction, and succession of the different forms of life which have left in the rocks a trace of their former presence as constituting its inhabitants. The considerations last urged have an especial interest from the point of view of vegetable paleontology, which presents a close parallel, though at a considerably lower horizon.

In the next annual report Dr. White goes over the same ground and sets forth his views anew, supported by fresh facts. In fixing the boundaries of the Laramie sea, he says (p. 49): "The geographical limits of the Laramie group are not yet fully known, but strata bearing its characteristic invertebrate fossils have been found at various localities within a great area, whose northern limit is within the British Possessions and whose southern limit is not further north than Southern Utah and Northern New Mexico. Its western limit, so far as known, may be stated as approximately upon the meridian of the Wahsatch range of mountains, but extending as far to the southwestward as the southwest corner of Utah, and its eastern limit is far out on the great plains, east of the Rocky Mountains, where it is covered from view by late formations and the prevailing *débris* of the plains. These limits indicate for the ancient Laramie sea a length of about one thousand miles north and south, and a maximum width of not less than five hundred miles. Its real dimensions were no doubt greater than those here indicated, especially its length; and we may safely assume that this great brackish-water sea had an area of not less than fifty thousand square miles."

He reiterates his statement that " With the exception of one species of *Axinœa*, one of *Nuculana*, and one or two of *Odontobasis*, no species usually regarded as of marine types have been found in any of the strata of the Laramie group," and pronounces all statements in conflict with this, even though made by himself, as the result of errors in stratigraphical determination. He also repeats the remark (p. 51) that " among all the invertebrate fossils which have yet been discovered in the strata of the Laramie group, none of the types are distinctively characteristic of the Cretaceous period according to any hitherto recognized standard," and he adduces a mass of facts in support of the view previously advocated, " that the Laramie is really a transitional group between the Cretaceous beneath and the Tertiary above (p. 52)."

In the sixth volume of Prof. Oswald Heer's great work on the Arctic fossil flora,[1] the eminent Swiss paleontologist approaches this question of the age of the American plant-bearing beds. As might be expected, he strongly defends Mr. Lesquereux's position as to the Eocene age of the Laramie group against the arguments of those who would refer it to the Cretaceous. He characterizes the doctrine that the Dinosaurs became extinct at the close of the Mesozoic as a " dogma," and, speaking of Cope's *Agathaumas*, says that it by no means proves that a Tertiary flora was contemporary with a Cretaceous fauna, " for a single animal does not make a fauna any more than one plant makes a flora," and instances the animal forms also found by Cope and others at the same horizon, which agree better with the Eocene faunas of France.

In the supplement to the third volume of the reports of Lieutenant Wheeler's Survey, which bears date 1881, Mr. John J. Stevenson again discusses the age of the Laramie group, adhering as warmly as ever to his previous views. As in his former reports, notwithstanding frequent denials in the meantime, he still insists (p. 154) that " farther north in Colorado characteristic Fox Hills fossils were obtained in abundance near the *summit of the fully recognized Laramie*." This and the farther statement (p. 154) that " the fauna is either marine or brackish-water" are both contrary to the definition of the Laramie group as laid down by Dr. White, and indicate that this geologist had been unable to distinguish the marine from the brackish-water strata. In his final conclusion that the Laramie merely constitutes the upper part of the Fox Hills group (p. 158), Mr. Stevenson seems to be sustained by no other authority, even the stratigraphical geologists, fully aware of the conformity of the deposition, not being willing to regard a marine and a brackish-water deposit as a single homogeneous group.

The Third Annual Report of the United States Geological Survey, published in 1883, contains Dr. White's " Review of the non-marine fossil mollusca of North America," illustrated by 32 plates, 22 of which are devoted to species of the Laramie group, all of which are described

---

[1] Beiträge zur miocenen Flora von Nord-Canada. Zürich, 1880, pp. 6-10.

in the text, and which furnish a thorough and complete account of the invertebrate fauna of that group. In the "Introductory remarks" which precede and the "General discussion" that follows this "Annotated Catalogue," Dr. White again sets forth his views upon this great series of rocks, which, however, had undergone no change. Although he now drops the term *Post Cretaceous*, he still regards the Laramie group "as a transitional group between the Cretaceous and Tertiary series, and therefore as representing a period partaking of both the Mesozoic and Cenozoic ages." In defining the group anew, he says that "the 'Judith River group,' 'Fort Union group,' 'Lignitic group,' 'Bitter Creek coal series,' 'Point of Rocks group,' and 'Bear River estuary beds,' are all parts of the great Laramie group," but that "the 'Wahsatch group,' 'Vermilion Creek group,' and 'Bitter Creek group' are regarded as at least approximately equivalent strata, constituting the oldest member of the purely fresh water Eocene Tertiary series of deposits in the West."

The most important part of this paper is the acute and suggestive geognostico biological discussion it contains respecting the origin and evolution of these brackish- and fresh-water invertebrate forms, but this is outside of our present limits, and need only be referred to.

The appearance of Prof. Archibald Geikie's new Text-Book of Geology, containing allusions to western American deposits, called forth from Dr. White a vigorous protest in his article on "Late Observations concerning the Molluscan Fauna, and the Geographical extent of the Laramie Group," in the American Journal of Science for March, 1883, in which he pronounces some of these statements erroneous, and says: "I do not hesitate to assert that not one of the molluscan species mentioned in that statement was ever found in strata of the Laramie group, the non-marine forms which he mentions being evidently those which were discovered by Mr. Meek in an estuary deposit of true Cretaceous age, at Coalville, Utah. Furthermore, not one of the numerous species which do characterize that group are anywhere mentioned in the book;" and, referring to Mr. Stevenson's writings, he says in the same article: "That any true Laramie strata ever alternate with those of the Fox Hills group, or any other marine Cretaceous group, or that any true marine fossils were ever collected from any strata of the Laramie group, I cannot admit. I regard all such statements as the result of a misunderstanding of the stratigraphical geology of the region in which such observations are said to have been made."

Having received a collection of typical Laramie fossils from the State of Nuevo Leon, Mexico, Dr. White is now able to extend the southern limit of the Laramie group to that point, and he states that the facts "show more and more clearly the integrity of the molluscan fauna of the great ancient intra-continental sea in which the Laramie group was deposited, and its separateness from the faunae of all other North American groups of strata (op. cit., p. 209)."

The latest utterance of this protracted debate is that of Mr. Lesque-

reux, in his new work just issued from the press on the "Cretaceous and
Tertiary Floras of the Western Territories."[1] He here consents, in
harmony with the general tendency of the time, to drop the term Eocene
from the title of this chapter and treat simply of the "Flora of the
Laramie group," without, however, surrendering his conviction that that
group belongs to Eocene time, which he reasserts, although he now
admits that "the flora of the Laramie group has a relation, remarkably
well defined, with that of Sézanne," to the east of Paris, where the plant
bearing travertines of the Lac de Rilly yield, according to the Marquis
Saporta, the oldest Tertiary flora yet discovered. He reviews the recently expressed views of White, Cope, and others, and seems quite
well satisfied with the state of opinion at the date of writing with respect to the age of the Laramie group.

NATURE AND EXTENT OF THE LARAMIE GROUP.

In the foregoing review of opinion I have sought to illustrate the
history of our knowledge of this remarkable formation of American
rocks, and to show how, as that knowledge increased, the wide fluctuations which characterized the period of general ignorance and limited
information gave way to a gradual convergence of views, an equilibra-
tion, as it were, of ideas, which is still going on and tending steadily
toward the final settlement of opinion in harmony with all the facts.

I have given special prominence to the evidence furnished by animal
remains and by stratigraphy, purposely leaving that from vegetable
remains, generally consistent with itself, undiscussed, because they form
the principal subject of this paper and can better be treated by themselves in a future place and in connection with other problems of greater
real importance than that of their geological age.

One of the advantages of the historical method here employed is that
it obviates the necessity of offering any special description of the group
under consideration as introductory to the treatment of its flora, the
reader being now much better prepared to understand such treatment
than any preliminary explanations of my own could have rendered him.

He perceives, from what has been said, that the Laramie group is an
extensive brackish-water deposit situated on both sides of the Rocky
Mountains and extending from Mexico far into the British North
American territory, having a breadth of hundreds of miles and representing some 4,000 feet thickness of strata. He can readily see that
when this deposit was made an immense inland sea must have existed
whose waters occupied the territory now covered by the Rocky Mountains. These waters were partially cut off from the ocean by intervening
land areas, through which, however, one or more outlets existed com-
municating with the open sea at that time occupying the territory of

[1] Report of the United States Geological Survey of the Territories (Hayden), Vol.
VIII, 1883, pp. 109-114.

the Lower Mississippi and Lower Rio Grande Valleys. That this great
inland sea spread over this entire territory is not at all disproved by
the absence of Laramie strata from large parts of it, since these parts
are situated, in most cases, in mountainous regions where the upper
strata might be expected to have been generally eroded away.

This Laramie sea existed during an immense period of time and was
finally but very gradually drained by the elevation of its bed, through
nearly the middle of which longitudinally the Rocky Mountains and
Black Hills now run. The exceeding slowness of this event is shown
by the fact, so clearly brought out by Dr. White, that the marine forms
of the Fox Hills strata, as they gradually found themselves surrounded
by a less and less saline medium on the rising of the intervening land
area, had time to become transformed and adapted to brackish-water
existence, while these new-formed brackish-water species, when the sea
at length became a chain of fresh-water lakes, had time again to take
on the characters necessary to fresh water life.

Dr. White recognizes the fact that the upheaval of the strata that
formed the bottom of this sea took place, not in one uniform process of elevation, but in a prolonged series of rhythmic fluctuations of level, whose
algebraic sum constituted at length a mountain uplift. But the numerous coal seams one above another that characterize the greater part of
these beds, and equally the successive deposits of vegetable remains at
different horizons, speak even more eloquently than any animal remains
can do of the oscillatory history of the bed of this sheet of water.

There may have been, and doubtless were, as Major Powell believed,
many islands scattered over the surface of this sea in Laramie time, and
the evidence generally warrants us in assuming that a low, level country
surrounded the sea, with marshy and swampy tracts. The islands and
shores were heavily wooded with timber that can be as certainly known
in its general character as we can know the timber of our present forests. But that for the greater part of the Laramie period there also
existed at no great distance a large amount of elevated land, there can
be no doubt. The deposits are chiefly siliceous in the southern districts
and argillaceous in the northern, but the nature of their deposition
points unmistakably to the existence of large and turbulent rivers that
fell into the quiet sea and brought down from areas of rapid erosion
immense quantities of silt corresponding to the nature of the country
over which they flowed in their course. Where these elevated sources
of this abundant detritus were then located is one of the great problems
for the present and the future geologist to work out.

The deposition of this material was almost always quiet, the particles
suspended in the turbid waters of the streams silently settling from
the buoyant waters of the sea as fast as they became distributed about
the mouths of the rivers, and thus embedding the leaves that periodically
fell in vast numbers into it. The marked absence of fruits, stems, and
other objects that possess considerable thickness shows that this was

the case, and also affords a rude index to the rate of deposition, since only such objects could be preserved as succeeded in being covered up. Thus by ascertaining the average rate of decay of vegetable substances and noting the objects of maximum thickness which are found preserved, the time necessary to form a deposit of that thickness becomes approximately known.

The discussions with regard to the age of the Laramie group which have been rapidly passed in review have, perhaps, sufficiently shown that it is impossible to refer that group either to the Cretaceous or to the Tertiary and in so doing harmonize all the facts that the group presents with those in conformity with which other deposits in other countries of the world have been so referred; but they have also sufficiently shown that this is not the fault of the investigators, but, so to speak, of the facts, and that the real disagreement is in the organic forms and the nature of the deposits, so that omniscience itself could never harmonize them with all kinds of forms and deposits in all parts of the world. It is, therefore, futile, and indeed puerile, longer to discuss this question, and we can well afford to dismiss it altogether and settle down to the more serious study of the real problems which still lie before us.

One of these problems is often confounded with the question of age, which should be rigidly distinguished from it. This is the question of synchronism. If it could be satisfactorily proved that the Laramie group was deposited at the same absolute time as the iron sands of Aix la-Chapelle, the Credneria beds of Blankenburg, or the travertines of Sézanne, this would indeed be a great gain to science. But as the animal and vegetable remains cannot be made to agree, it seems hopeless to attempt to arrive at complete harmony in this respect. The most that can be profitably undertaken is to find two or more deposits widely separated geographically in which either the floras, the invertebrate faunas, or the vertebrate faunas substantially agree. With regard to the invertebrate faunas this seems hopeless so far as the Laramie group is concerned. If that group was deposited in the manner above described, it would be difficult to find another which owed its existence to identical conditions; and if this state of things has occurred at more than one point upon the globe, the chances are again greatly diminished for it to have occurred at the same period of geologic time. But even supposing such a combination of coincidences possible, if the Laramie forms are the modified descendants of antecedent marine forms, there is no probability that the conditions at any other point on the earth's surface could be so nearly identical with those obtaining there that precisely the same modifications would take place to adapt the marine forms to the brackish-water habitat. The chances are therefore infinity to one against the existence of other beds that shall contain an invertebrate fauna identical with that of the Laramie group.

It is therefore truly surprising to learn that "several of the species found in the brackish-water layers at the base of the Bitter Creek group are closely related to species found in similar deposits in Slavonia and referred to the Eocene Tertiary by Brusina."[1]

With regard to vertebrate remains, this objection does not apply, and could they be made to harmonize with themselves they might, perhaps, be trusted to some extent as indices of synchronism in widely separated localities. But, as shown by Cope, they do not thus agree, for the Laramie forms include genera that are regarded as characteristic of Cretaceous and others that are regarded as characteristic of Tertiary strata. This should surprise no one. The law that has been laid down by paleontologists, that the same epochs in geologic time produced the same living forms — which is the converse of the assumption commonly acted upon, that the occurrence of the same forms proves the beds containing them to be of the same age — is contrary to the now well established principles of geographical distribution, according to which the earth is subdivided into a large number of faunal areas more or less clearly marked off one from another. The peculiarity of this principle which is of most importance to paleontology is that these territorial subdivisions represent faunas not merely different from one another, but showing different degrees of biologic development as development is supposed to have gone on in the animal kingdom. Every one knows that the fauna of Australia belongs to an undeveloped type, being marsupial in aspect so far as its mammals are concerned. The types of South America are lower than those of North America, and the latter lower than those of Asia and Europe. If all the present faunas of the globe were buried under its soil it is clear that it would not only be impossible to harmonize the deposits of different continents, but that the inference now freely drawn by paleontologists that the less developed forms demonstrate their existence at earlier epochs would lead to grave mistakes and be generally false. New Zealand is now in its age of birds, while the Galapagos Islands are still in that of reptiles, or the Mesozoic age.

VEGETATION OF THE LARAMIE AGE.

Confining ourselves, then, for the future to the other kind of land life and the only remaining form of life, that of plants, we may look at the question of synchronism by the light of this class of data from the same general point of view as we have done by the light of the two kinds of animal life which we have just considered. And, first, what ought we to expect the flora of the Laramie group to teach respecting the synchronism of its deposits with those of other parts of the world? Clearly, as in the land vertebrate life, there is no special obstacle to this form of inquiry, such as the invertebrate aquatic life presents, arising

---

Dr. White, in "Geology of the Uinta Mountains," p. 86.

out of the manner in which the Laramie sea was produced and the
changing constituents of its waters. But all the other difficulties present
themselves here as in the case last considered. While the vegetable
remains seem to be more harmonious in pointing to a somewhat
later period of time for their deposition than do those of vertebrate
animals, the impropriety of inferring absolute synchronism from substantial
agreement of forms is here even greater than in the other case.
Taking the present flora of the globe as a criterion, we find that the
geographical distribution of plants is more uneven than that of animals.
Floral realms are more numerous and distinct than faunal realms, and
the more serious obstacle that some areas furnish types representing
less developed floras than others exists here as in the case of animals.
The Proteaceous and Myrtaceous flora of Australia may be regarded as
rudely corresponding to its marsupial fauna.

It is true that the paleontological doctrine of synchronism already
stated is supported, as against the facts of geographical distribution,
by the well established principle that older faunas and floras were characterized
by less variety and greater uniformity of distribution over
the earth's surface, which is verified in a remarkable manner by the
well known uniformity of the flora of the Carboniferous epoch at all
points where it has been discovered. And Baron Ettingshausen has
shown that this principle continued in operation down to the close of
the Tertiary age, though, of course, in a reduced degree, so that the
present extraordinary variety in the floras of different countries must
be largely attributed to the agency of the successive glacial epochs
which have occurred since Tertiary time in driving the floras southward
and out on the southern plains to be destroyed on the return of
warmer climatic influences or compelled to intrench themselves upon
the summits of the mountain ranges, while new and constantly varying
forms became developed to take their places in the lowlands. Still,
the uniformitarian law, that in its more general aspects the phenomena
taking place on the earth in past geologic ages were the same as those
which are still taking place, forbids us to assume that even as far back
as Laramie time the same or any very similar flora occupied different
hemispheres of the globe.

This much, however, can be said in favor of the flora of the Laramie
group as affording data for the study of its deposits: that its remains
occur far more abundantly than do those of any of the other forms of
life. The low forest-clad shores and islands of the Laramie sea, which
probably extended back at many points into extensive lagoons and vast
swamps, were peculiarly adapted for receiving, as its muddy waters
were for embedding, the various kinds of vegetable matter that found
their way into them. The swamps formed extensive beds of peat, and
vast marshes densely covered with cane, bamboo, and scouring rush
left thick annual accumulations of vegetable matter which, at points of
slow temporary subsidence, formed the coal beds. The plant beds which

usually overlie these coal beds tell us that the rate of subsidence had now exceeded that of the growth of the deposit and the shallow sea had gained access, burying the last of the plants under its siliceous or argillaceous precipitations where they were preserved. Almost everywhere, even when no leaves or twigs are present, we find the stout subterranean rhizomas of the cane and the scouring rush, which, not having to be covered up, stood a far better chance to be preserved. But in numberless places the profusion of leaves is so great that there is too little rock between them to render it easy or even possible to separate them and obtain complete specimens. Above the plant beds, and occupying the intermediate strata between these more carbonaceous deposits of coal, reeds, and leaves, we find thicker and often massive beds of sandstone or marl, which seem to denote the presence over the former deposits of deep water produced by continued subsidence and the recession of the shore lines to distances too great for the access of the falling leaves, and the continuance of these conditions through prolonged periods of time.

If now we compare the flora of the great Laramie group, as thus described, with its invertebrate fauna, as elaborated by Dr. White, we find that in its *ensemble* the former is much more variable than the latter. The dicotyledonous species differ greatly at different parts of the area covered by the rocks of this group, so greatly, indeed, that it is not surprising that both Mr. Lesquereux and Dr. Newberry regard the Fort Union plants as belonging to a different age from those of the Wyoming and Colorado Laramie. Still, as I shall endeavor to show, this difference is not so great as it at first appears, and not sufficient to warrant this conclusion. In the first place, this difference appears chiefly in the dicotyledonous species, the only marked exception being that palms occur much more abundantly in the southern than in the northern districts. The same forms of reed-like plants are common at all points, while the Coniferæ do not differ more than might be expected on the theory of synchronism. The same is true of the abundant Equisetums, while very few ferns are found within the group.

Aside from the presence of palms the flora of the lower districts indicates a difference of climate greater than can be accounted for by the small difference of latitude. This is proved by the great prevalence of the genus Ficus and the presence of Cinnamomum, both of which are rare or wanting in the Fort Union group, while in the latter occur a great variety of Populus common to cold climates and the genus Corylus in abundance, absent from the Wyoming and Colorado beds. There are two ways in which these differences may be explained, or at least an explanation of them attempted, without denying the great difference of climate. In the first place, it is probable that the more southern parts of the Laramie sea were also much nearer the ocean on both the east and the west sides, and hence enjoyed a more equable climate, as well as one more moist, such that few of the trees and shrubs would lose their leaves by the action of frosts and that subtropical species, like the palms, the figs, and the cinnamons, could subsist. In the second place, it must be remembered that the Laramie period was a very prolonged one, and within it there was time for considerable alteration of climate on this continent or even on the whole globe. But even admitting that this was too slight to be perceptible, the changes that took place in the form of the continent and the distribution of land and water on it during that time might have been sufficient to produce marked effects and render the later floras of the Laramie age quite different from its earlier floras.

The Fort Union beds, containing the genera Corylus, Sapindus, and other forms of recent aspect not found in the Bitter Creek and Golden deposits, are believed to be high up in the series; and I have myself found and explored others within the general district included by that group which I have proved stratigraphically to occupy a considerably lower horizon, and in which these forms of recent aspect not only do not occur, but some of the most characteristic Laramie types, such as *Trapa microphylla* and *Pistia corrugata*, do occur, together with other forms not previously known as Laramie. In fact, it is well known that the Fort Union Laramie is everywhere thinner than the more southern deposits, none of the sections making it over 3,000 feet in thickness. The beds to which I refer rest immediately upon the typical Fox Hills, and therefore represent the lowest strata present in that section. I am not yet prepared to speak upon the precise affinities of this lower Fort Union flora, not having completed the elaboration of my material, but I can say this much, that besides containing some of the more southern Laramie forms, its general aspect indicates a much warmer climate than that which prevailed at the time of the deposition of the Corylus and Viburnum beds above.

Fully conceding, as I do, that the geological age of the Laramie group cannot, for the reasons stated, be proved by its flora alone, and holding that even great similarity of flora would not be conclusive as to synchronism of deposit. I have still thought it instructive, in view of the warmth with which the Cretaceous and Tertiary theories for the age of this group have been respectively advocated, to make some general comparisons of its flora with those of the extreme upper Cretaceous and lower Tertiary of those parts of the world where the stratigraphical position has been settled. In the several elaborate tables of distribution of the species of the Laramie group which Mr. Lesquereux has drawn up and employed to demonstrate its Eocene age, it is noticeable that he has seemed to ignore almost altogether the existence of a large upper Cretaceous flora lying entirely above the Cenomanian and its American equivalent, the Dakota group. In a paper which appeared in the American Journal of Science for April, 1884, I succeeded in getting together 260 species of Dicotyledons alone from this formation, which I designated as Senonian, and in a table published in the last Annual Report of the Geological Survey (1883-'84, p. 440) I showed that 354 Senonian species were then known, a flora slightly larger than that of the Laramie group. The principal localities from which this flora is derived are: the Iron sands of Aix-la-Chapelle, the Credneria beds of Blankenburg and Quedlinburg in the Harz Mountains, numerous deposits in Westphalia, the Gosau formation in Austria, the Lignites of Fuveau in Provence, France, the beds of Patoot, Greenland, and those of the Peace and Pine Rivers, British America, and of Vancouver and Orcas Islands on the Pacific coast. All of these beds are quite definitely fixed in the upper Cretaceous, those of Europe being well known. As regards the others, Professor Heer states that those of Patoot possess a molluscan fauna identical with that of the Fox Hills group of North America, and Mr. G. M. Dawson correlates those of the interior of British America with the Niobrara of Meek and Hayden, and those of the Pacific coast with the Fox Hills. All authorities agree, however, that all these beds are lower than the Laramie, and Dawson makes our Fox Hills the equivalent of the Maestricht and Faxoe beds, the white chalk, Danian, or extreme upper Cretaceous of Europe.

EXPLANATION OF THE TABLE OF DISTRIBUTION.

The following table aims to give all the fossil plants which have been thus far authentically described and recorded (1) in the Laramie group as above defined, (2) in the Senonian as last described, and (3) from the beds that have been unanimously referred to the Eocene. This last naturally excludes the Green River group, which is regarded as the American Eocene of the West by nearly all authorities except Mr. Lesquereux. As this one prominent author assigns the Laramie group (as defined by him) to the Eocene and places the Green River deposits in a higher formation, and as it is chiefly to test this question that the table and its discussion are intended, it would manifestly vitiate the argument to prejudge the question by adding the Green River group to the accepted Eocene.

In preparing this extensive table it has been my aim to embody in it as large an amount of information bearing not only upon the age and synchronism of the Laramie group but also upon all the collateral problems arising out of a study of the flora of that group as could be condensed into that amount of space. The plants are systematically arranged according to the latest botanical classifications, the names of the subordinate groups being entered in their proper places and distinguished by different type. The genera occupy separate lines and the number of species represented in each genus is given in each column on those lines, the occurrence of species in the several formations being denoted by the customary sign (+) employed by most authors for this object.

In the vertical arrangement the Laramie group is placed first merely because it is the group under immediate consideration, the Senonian next, because lowest, and because it is to its flora that it is especially desired to direct attention; the Eocene properly coming last. The first subdivision of the Laramie is intended to cover all the beds recognized by Mr. Lesquereux as belonging to that group. The Carbon and Evanston coal beds, excluded by him, follow, the two columns covering all the plants from the central and southern areas, the third being reserved for those of the northern districts, generally included under the name of Fort Union group. To this latter group, as undoubtedly belonging to a still more northern extension of it, I have assigned the species named by Sir J. W. Dawson,[1] as having been found in the Laramie of the British Provinces. These I have distinguished by the letters B. A. and the frequent coincidence of these letters with the regular sign for the species sufficiently attests the correctness of this conclusion. Most of the interrogation points occurring in this column represent cases where the fossils have been reported from the localities denominated "Six miles above Spring Cañon, near Fort Ellis, Montana." "Yellowstone Lake," "Elk Creek," and "Snake River." These plants are all classed by Mr. Lesquereux in his first and lowest group, or true Laramie, but upon careful investigation I am tolerably well satisfied that they belong to the Fort Union deposits. Their northern position and the known fact that these deposits extend far up the Yellowstone and Missouri Rivers would naturally favor this view, but it is the internal evidence afforded by the species themselves which is most convincing. A large proportion of the forms from this locality are also found in the true Fort Union beds and among these occurs *Platanus nobilis*, otherwise wholly characteristic of these beds. It is true that one species of Ficus and one palm occur here, but the genus Ficus is no longer excluded from the Fort Union group, while the occurrence of palms in that group has been recognized from the first.

The several acknowledged upper Cretaceous beds enumerated on a previous page are each given a separate column, and five of the most characteristic Eocene localities are also thus distinguished, the sixth column being devoted to several less important and some outlying beds referred to that age. In the last column the several localities which have been set off by some authors from the true Eocene and classed as Paleocene are grouped together. The principal beds of this class are the Travertines of the Lac de Rilly near Sézanne, to the east of Paris; the supra-lignitic deposits about Soissons, the "Sables de Bracheux;" and the so-called "Marnes Heersiennes" of Gelinden, all situated in Northern France and adjacent Belgian territory and immediately joining the only slightly lower Maestricht deposits.

The three broader columns which complete the body of the table

[1] On the Cretaceous and Tertiary Floras of British Columbia and the Northwest Territory. Transactions of the Royal Society of Canada, 1883, pp. 15-34, Pl. I-VIII (see list of Laramie plants on page 32).

merely sum up the data contained in these more detailed entries and exhibit the three formations side by side in compact form for ready comparison.

To this are added eleven columns for the purpose of indicating the vertical range of both the genera and the species. The first of these, in which the letter referring to the foot-note is substituted for the conventional sign, shows those forms which occur below the Cretaceous, the foot-notes showing the formations in which found. The headings of the other ten columns sufficiently explain themselves.

The geographical distribution of living genera, so far as practicable, and of genera closely allied to extinct ones, is also given in foot-notes, and the number of species of living phenogamous genera, as estimated by the highest botanical authorities, is indicated by figures in parenthesis. The importance and significance of this feature will be discussed in the proper place.

Table of distribution of Laramie, Senonian, and Eocene plants.

Species represented.	Laramie.			Senonian.								Eocene.						Summary of the foregoing.			Other formations in which found.										
	Bitter Creek, Golden, Raton Mountain, &c.	Carbon and Evanston.	Fort Union group.	Aix-la-Chapelle.	Harz District.	Westphalia.	Gosau formation, Austria.	Lignites of Fuveau, Provence.	Patoot, Greenland.	Peace and Pine Rivers, British America.	Vancouver and Orcas Islands.	Paris Basin.	Aix in Provence.	Arkoses de Brives.	London clay.	Mounts Bolca, Pastello, and Promina.	Other typical Eocene.	Paleocene (Bracheux, Sézanne, Soissons, Gel.).	Laramie.	Senonian.	Eocene.	Lower than Cretaceous.	Lower Cretaceous (below Cenomanian).	Cenomanian.	Dakota group.	Green River group.	Oligocene.	Miocene.	Pliocene.	Quaternary.	Living species.	Genera extinct.
Series I.—CRYPTOGAMIA.																																
CLASS I.—CELLULARES.																																
Fungi:																																
Sphæria, Hall	3								1				2						3	1	2					+		+				
cretacea, Heer									+											+												
lapidea, Lx	+																		+													
minutula, Sap													+								+											
myricæ, Lx	+																		+		+					+						
proxima, Sap													+								+											
rhytismoides, Lx	+																		+													
Sclerotium, Tode	1																		1									+				
rubellum, Lx	+																		+													
Lichenes:																																
Opegrapha, Ach	1																		1													
antiqua, Lx	+																		+													
Algæ:																																
Confervites, Schp				1																1		(a)					+	+				+
Aquensis, Deb. & Ett				+																+												
Thorea, Bory																3					3											
Brongniartii, M																+					+											
intermedia, M																+					+											
Janii, M																+					+											
Monemites, M																1					1											+
species, M																+					+											
Caulerpa, Lam *b*																	3				3											
annulata, Schp																	+				+											
arbuscula, Schp																	+				+											
arcuata, Schp																	+				+											

a Wealden.

b Tropical seas.

Table of distribution of Laramie, Senonian, and Eocene plants—Continued.

a Silurian, Devonian   b Lias, Coral.   c Muschelkalk, Lias, Oolite, Coral.

Table of distribution of Laramie, Senonian, and Eocene plants—Continued.

Species represented.	Laramie.			Senonian.								Eocene.										Summary of the foregoing.				Other formations in which found.											
	Bitter Creek, Golden, Elko, Point of Rocks, Raton Mountain, &c.	Carbon and Evanston.	Fort Union group.	Aix-la-Chapelle.	Harz District.	Westphalia.	Gosau formation, Austria.	Lignites of Province.	Lignites of Proveau.	Patoot, Greenland.	Pecten and Pine Rivers, British America.	Vancouver and Great Island.	Paris basin.	Aix in Provence.	Arkona de Drivers.	London clay.	Monnts Bolca, Pas.	Radoboj and Promina.	Other typical Eocene.	Pabeorene (Eratz)lux, Szeząu, Pelaisuu,(Gel.)	Laramie.	Senonian.	Eocene.	Lower than Creta-ceous.	Lower (Neocomian).	Lower'(Cretaceous)(below Cenomanian).	Cenomanian.	Dakota group.	Green River group.	Oligocene.	Miocene.	Pliocene.	Quaternary.	Living species.	Genera extinct.
Chondriten, St.—Continued.																							+													+
subverticillata, Presl																							+			+										+
Targionii, St							+																+													+
Targionii arbuscula, Hr-r																							+													
Targionii expansum, Heer																							+													
Zinnwaldi, M							+																+			+										+
Dobrovria, LamX																							+													+
Agardhiana, Schp																							+													
Dobrovria, M																							+			+										+
canbroensis, Schp																							+													
flabelliformis Schp																							+													
intra, Lx																							+													
Gazolaum, Schp																							+													
Sambrana, M																							+													
sphaerococcoides, Schp																							+													
Thuresenii (Brosni,) Schp																							+													
Tasulium, Heer																							+													+
abelaides, Hos. & Mck																							+													
Fischeri, Heer																							+													
Protophyree, M																							+													
speetabilis, M																							+													
Cerunites, M																							+													+
species, M																							+													
Palmia, M																							+													
species, (a)																							+													+
Confhns, J																							+													+
(!) Michelotti Schp																							+													
Halimerites, Schp																							+													
contortuplicatus, March																							+													+

a Silurian, Oolite.   b Devonian.

Table of distribution of Laramie, Senonian, and Eocene plants—Continued.

Table of distribution of Laramie, Senonian, and Eocene plants—Continued.

Cylindrites, Göpp.—Continued.
  ramicus, Hos. & Mrk.
  convallinus, F.O.
Laminaritcs, Sternb.
  articulatus, Wat.
  flabellaris, Wat.
  iridophyllus, M.
  Jovii, Wat.
  macrophyllus, M.
  quadralms, Wat.
  stipitatus, Wat.
Tænurius, Fr.?
  Brisacensis, Villa
Aristophyces, M.
  Agardhianus, M.
Nematolitcs, M.
  limuloides, M.
Characeæ:
  Chara, Vaill.?
  Archiaci, Wat.
  Brongniarti, Al. Br.
  depressa, Wat.
  Datempoei, Wat.
  Grepini, Hier.
  Kyperına, Sap.
  helicteres, Brongn.
  Lemani, Brongn.
  Lydelli, Al. Br.

a Silurian, Lias. *b* Lias, Oolite. *c* Throughout, temperate and tropical latitudes. *d* Oolite.

Table of distribution of Laramie, Senonian, and Eocene plants.—Continued.

Species represented.

Chara, Vaill.—Continued.
minima, Sap.
australis, Wat.
Spathacensis, Wat.

Muscineæ:
Marchantia, L.
gracilis, Sap.
Sezannensis, Sap.
Muscites, Brongn.
perteger, Sap.
redivivus, Sap.

CLASS II.—VASCULARES.

Filices:
Sphenopteris, Brongn.
elongata, Newby.
guyotii, Heer.
Lakesii, Lx.
membranacea, Lx.
nigricans, Lx.
Davallia, Sm.
tenuifolia, Sw.
Davallites, Pres.
Richardsoni, Daws.
Hymenophyllum, Knitl.
eoquisita, Brongn.
Xeuropteris, Brongn.

a Chiefly tropical.
b Subcarboniferous, Carboniferous, Permian, Rhetic, Lias, Oolite, Coral, Wealden.
c Carboniferous.
d America, Mexico and West Indies to Patagonia, India, and East India Islands, Australasia, Mascarene Islands, Saint Helena, Pacific Islands.
e Carboniferous.
f Subcarboniferous, Carboniferous, Permian, Keuper, Oolite.
Tropical Asia, Japan, Malay Archipelago, New Zealand.

Table of distribution of Laramie, Senonian, and Eocene plants — Continued.

Column headers (rotated):

Laramie
- Bitter Creek, Golden, Raton Mountain, &c.
- Carbon and Evanston
- Port Union group

Senonian
- Aix-la-Chapelle
- Harz District
- Westphalia
- Iowa formation, Amboy clays, Lignites of Fureau, Provence.
- Patoot, Greenland
- Lower and Upper British America, Vancouver and Orcas Islands
- Paris Basin
- Aix in Provence
- Aix-knees de Brivs

Eocene
- London clay
- Monnte Bolca, Pan-heim and Promina
- Other typical Eocene
- Paleocene (Sézanne), Sézanne, Soissons, (&c.)

- Laramie
- Senonian
- Eocene

Summary of the foregoing.
- Lower than Creta
- Lower Cretaceous (inc. Wealden), Neocomian (Wealden)
- Cenomanian
- Dakota group
- Green River group
- Other formations in which found:
  - Oligocene
  - Miocene
  - Pliocene
  - (Quaternary)
  - Living species
  - Genera extinct

Species represented.		
Neuropteris, Brongn.—Continued.		
Castor, Daws		
Nilssonia, Brongn		
lata, Daws		
Pecopteris, Brongn b		
Bohemica, Corda		
heteropbila, (Ung.) Schp.		
aratia, St.		
Zippei, Corda		
specioa, Daws		
Gomopteris, Presl		
Dalmatica, Al. Br		
polypodioides, Ett		
Stiriaca, Al. Br		
Danaites, Göpp		
Schlotheimi, Deb. & Ett.		
Bontreutatea, Deb. & Ett.		
cardinalis, Deb. & Ett.		
Zamiopteris, Deb. & Ett.		
Gepperti, Deb. & Ett.		
Beania, Deb. & Ett.		
Caloperti, Deb. & Ett.		
Raphcbia, Deb. & Ett.		
heteropteroides, Deb. & Ett.		
Pteridoleimma, Deb. & Ett.		
aminifolium, Deb. & Ett.		
Bilinquia, Deb. & Ett.		
Braineana, Deb. & Ett.		
deperditum, Deb. & Ett.		
dictyodes, Deb. & Ett.		

a Rhetic. b Subcarboniferous, Carboniferous, Permian, Kemper, Rhetic, Lias, Oolite, Wealden. c Carboniferous, Permian. d Carboniferous, Permian.

Table of distribution of Laramie, Senonian, and Eocene plants—Continued.

Species represented.	Laramie			Senonian						Eocene										Summary of the foregoing			Other formations in which found.											
	Bitter Creek, Golden, Elatton Mountain, &c.	Carbon and Evanston.	Fort Union group.	Aix-la-Chapelle.	Harz District.	Westphalia.	Gosau formation, Aus.	Lignites of Fuveau, Provence.	Patoot, Greenland.	Peace and Pine Rivers, British America.	Vancouver and Orcas Islands.	Paris Basin.	Aix in Provence.	Arkosses de Brives.	London clay.	Monnta Bolca, Pat-ella, and Promina.	Other typical Eocene.	Paleocene (Bracheux,).	Sezanne, Soissons,(fol.).	Laramie.	Senonian.	Eocene.	Lower than Trias.	...ceous.	Lower Cretaceous (the low Cretaceous).	(Lionmian,?)	Dakota group.	(Green River group)	Oligocene.	Miocene.	Pliocene.	Quaternary.	Living species.	(Genera extinct).

Pteridoleimma, Deb. & Ett.—Continued
dubium, Deb. & Ett
Elizabethae, Deb. & Ett.
gymnorachis, Deb. & Ett.
Hableizerri, Deb. & Ett.
Heeranum, Deb. & Ett.
Knlcanlarbi, Deb. & Ett.
Konnckmanum, Deb. & Ett.
leptophyllum, Deb. & Ett.
Michelini, Deb. & Ett.
odontopteroides, Deb. & Ett.
orthophyllum, Deb. & Ett.
pecopteroides, Deb. & Ett.
pseudodianthum, Deb. & Ett.
litzianum, Deb. & Ett.
Scortesi, Deb. & Ett.
Waterkeyni, Deb. & Ett.
species, Sap
Taenopteris Bongn.
affinis, M
desperdita, Heer.
Gibbsii, Newby
planosus, Daws
Oleandridium, Schp.
Johniuni (Wat.) Schp.
Michelni (Wat.) Schp.
obtusum (Wat.) Schp.
Carolopteris, Deb. & Ett.
Aquensis, Deb. & Ett.
asplenioides, Deb. & Ett.

a Permian. b Kterke, Oolite, Wealden.

Table of distribution of Laramie, Senonian, and Eocene plants — Continued.

Species represented.	Laramie			Senonian.								Eocene.								Summary of the foregoing.			Other formations in which found.											
	Bitter Creek, Golden, Union Mountain, &c.	Carbon and Evanston	Fort Union group.	Aix-la-Chapelle	Harz District.	Westphalia.	Gosau formation, Anatrim.	Lignites of Fuveau, Provence.	Patoot, Greenland.	Pence and Pine Rivers, British America, Vancouver and Great Islands.	Paris Basin.	Aix in Provence.	Arkesus de lettres.	London clay.	Monte Bolca, Provence.	Other typical Eocene.	Palœocene (Bracklux; Sezanne, Solnofer (fe).	Laramie.	Senonian.	Eocene.	Lower than Cretaceous.	Lower Cretaceous (the low Cenomanian).	Cenomanian.	Dakota group.	Green River group.	Oligocene.	Miocene.	Pliocene.	Quaternary.	Living species.	(Genera extinct.)

Monimisia, Deb. & Ett

Aquisgranensis, Deb. & Ett

poli pedioides, Deb. & Ett

Polypodium, L.d

Gelnslana, Sap.

Chrysodium, Fée.: Acrostichum, b?

Lauxccanum, Vis

Podoeum, Ett

affino, Ett. & Gard

lyeopodioides, Ett. & Gard

polypodioides, Ett. & Gard

Glossoclaura, Ett

transentinum, Ett. & Gard

Phlegopteris, Feo

Dumberti, Heer

Grothiana, Heer

Kotuerupi, Heer

praecuspidata, Ett. & Gard

Cladalluce, Swc

polinera, Sap.

Adiantum, L.d

apalophyllum, Nap

di mainorve. Heer

species, Ett. & Gard

Hewardia, Smith: (Adiantum, L.)

regia, Ett. & Gard

Adiantica, Gaup.

rassebrerodus, Deb. & Ett.

a Nearly cosmopolitan.

b Most tropical and warm regions; this section chiefly in old World.

c Temperate and tropical America, Asia, Mediterranean region, Indian and Malay Archipelagos, Africa, Canaries and Madeira, Bourbon.

d Nearly all temperate and tropical regions, chiefly in tropical South America.

e Subcarboniferous, Carboniferous.

Table of distribution of Laramie, Senonian, and Eocene plants—(Continued).

Species represented.	Laramie.				Senonian.													Eocene.					Summary of the foregoing.				Other formations in which found.												

Adianthes, Göpp.—Continued.
Decus-Brown., Deb. & Ett ...
lae-ruz, Sap ...
parlatorius, Daws ...
Sch-kohl, Ett ...
Vedenum, Sap ...
Blechnum, La ...
atarium, Sap ...
Brauni, Ett ...
Pteris, L.b ...
aquensis, Sap. & Gard ...
Bourrouse, Ett. & Gard ...
candigera, Sap ...
erosa, La ...
glossopteroides, Daws ...
Hookeri, Heer ...
Hantii, Ett ...
lunariiformis, Sap ...
longipennis, Heer ...
Prestwichii, Ett. & Gard ...
pseudoparschioides, La ...
Russelli, Newby ...
subsimplex, La ...
Asplenium, Lc ...
Broquianti, Deb. & Ett ...
cartagetonides, Deb. & Ett ...
carphurum, Sap ...

a America, Mexico, and Florida to Chiloe), Southern China, Himalayas, Malay Archipelago, temperate Australe, New Caledonia, Pacific Islands.　b Nearly all parts of the world, passing the Arctic Circle in Lapland.　e Nearly cosmopolitan, chiefly tropical and temperate.

Table of distribution of Laramie, Senonian, and Eocene plants—Continued.

Species represented.	Laramie.				Senonian.											Eocene.								Summary of the foregoing.				Other formation in which found.												
	Bitter Creek, Golden, Raton Mountain, &c.	Carbon and Evanston.	Fort Union group.	Aix-la-Chapelle.	Harz District.	Westphalia.	Gosau formation, Aus-tria	Lignites of Fuveau, Provence.	Patoot, Greenland.	Peux d'Or Rivers, British America.	Vancouver and Orcas Islands.	Paris Basin.	Aix in Provence.	Arkows de Brives.	London clay.	Monte Bolca, Tas-tello and Fronular.	Other typical Eocene.	Paleovène (Brixham, Sezanne, Soissons, &c.)	Laramie.	Senonian.	Eocene.	Lower than Creta-ceous.	Lower Cretaceous (the Jura Cenomanian).	Cenomanian.	Dakota group.	Green River group.	Oligocene.	Miocene.	Pliocene.	Quaternary.	Living species.	Genera extinct.				

Aspleniam, L.—Continued.
Fœrsteri, Heer, & Ett.
Pengelianum, Heer
scrobiculatum, Heer
tenerum, Lx
Wegmanni, Brongn
Aspidites, Göpp.
pradiosauridea, Ett. & Gaud
Menisphyllum, Ett. & Gaud
elegans, Ett. & Gaud
Aspidium, Sw e
Kraperlii, Newby
Œrsted, Heer
Reichianum, St.
Woodwardia, Sm b
latiloba, Lx
latiloba minor, Lx
Pyrenæa, Ett. & Gaud
Dicksonia, Heer
Groenlandica, Heer
Diplazium, Sw d
Millefii, Heer
Lastrea, Presl
Goldiana, Lx
intermedia, Lx
Gymnogramme, Desv f

a Temperate and tropical North America, Southern Asia, Madeiras, and Canaries. b Temperate and tropical North America, Southern Europe, Pacific Islands, Madeiras, Azores, Bourbon, Asia, Australasia, East India Islands, South Africa, New Hebrides. c Tropical America, temperate North America, Southern Europe, Pacific Islands, Madeiras, Azores, Bourbon, Asia, Australasia, East India Islands, South Africa, New Hebrides. d Southern Asia. Himalayas, Japan, India and Malay Archipelagos, Angola, Mascarene Islands, Pacific islands, tropical America, Mexico, West Indies. e Nearly cosmopolitan. f All tropical regions, west coast North America to Vancouver Island, Cape Colony, Japan, Himalayas. g Nearly cosmopolitan. Species sometimes referred to Aspidium and Nephrodium.

Table of distribution of Laramie, Senonian, and Eocene plants—Continued.

Table of distribution of Laramie, Senonian, and Eocene plants.—Continued.

Table of distribution of Laramie, Senonian, and Eocene plants—Continued.

Table of distribution of Laramie, Senonian, and Eocene plants—Continued.

Series II.—PHÆNOGAMIA.

CLASS I.—GYMNOSPERMÆ.

Cycadaceæ:
 Zamites, Brongn.
  acutila.us, Sap.
  species, Sap.
 Zamiostrobus, Endl.
  gibbus (Reuss) Schip.
  mirabilis, Lx.
 Dioonites, Born.d.
  borealis, Daws.
 Pterophyllum, Brongn.
  Rumeziana, Stiehl.
 Cycadites, Krauss.
  Ungin, Daws.
 Cycadeoxylon
  Westfalicum, Hos. & Mek.

Coniferæ:
 Abietites, Göpp.a.
  curvifolius, Dunk.
  dubius, Lx.
  setiger, Lx.
 Tsuga, Carr.
  Rostri, Ett. & Gard.
 Pinus, Lf.

a (Zamia 20.) Tropical and subtropical North America.
b Lias, Oolite, Coral, Wealden.
c Oolite, Coral, Wealden.
d (Dioon 2.) Mexico.
e Pernhan, Keuper, Rhætic, Lias, Oolite, Wealden.

f Wealden.
g (5) 2 Asia. 3 North America.
h (20) Extratropical, North America. Very few tropical Eastern Asia, West Indies, and Central America.

f Keuper, Rhætic, Oolite, Wealden.
g Carboniferous, Rhætic, Lias, Oolite, Wealden.
h (Abies 18.) Extratropical, Northern Hemisphere.

Table of distribution of Laramie, Senonian, and Eocene plants—Continued.

Pinus. L.—Continued.
Aspensis, Sap.
Baifgi, Ett. & Gard.
Bovtichiauki (Carr.) Ett. & Gard.
Coquandi, Sap.
Defranceri, Brongn.
diversifolia, Sap.
Dikeni (Bow.) Ett. & Gard.
gracilis, Sap.
immilis, Sap.
macrocephalus (L. & H.) Ett. &
    Gard.
monstrocephala, Row. & Mck.
evata, L. & H.
Platonis, Italy
Proctwichii, Ett. & Gard.
robustiflata, Sap.
Sequoiasomii, Wat.
Araucaria, Juss. a
Gippert, St.
Araucarites, St.
immediately, L. & H.
Hartigi, Dunk.
Agalis, Salcb. (Dammara. Lamb).
    macrosperma, Heer.
    microlepis, Heer.
Cuminghamia, St.
Proctwichia, Ett. & Mck.
squamosa, Heer.

a (1b) South America, Australia, New Caledonia, and South Pacific islands.   c (8-10) Malay Archipelago, Fiji Islands, New Caledonia, New Zealand, and tropical East Australia.
b India, Wealden.   d (Cunninghamia 1.)   Japan and China

Table of distribution of Laramie, Senonian, and Eocene plants.—Continued.

Species represented.	Laramie		Senonian								Eocene							Summary of the foregoing.				Other formations in which found.											
	Bitter Creek, Golden, Raton Mountain, &c.	Carbon and Evanston.	Fort Union group.	Aix-la-Chapelle.	Harz District.	Westphalia.	(Gosau formation, Austria.)	Lignites of Provence.	Patoot, Greenland.	Preuss and Pine Rivers, British America.	Vancouver and Oregon.	Tabanda.	Paris Bassin.	Aix in Provence.	Atkasses de Brives.	London clay.	Sezanne, Belgium, Prov.	Memnin Bacin, Pas, Sandstello, and Premina.	(other typical Eocene).	Paleocene (Bracheux, Sezanne, Soissons, &c.)	Taranic.	Senonian.	Eocene.	Lower than Trias.	Lower Cretaceous (below Cenomanian).	Cenomanian.	Dakota group.	Green River group.	Oligocene.	Miocene.	Pliocene.	Quaternary.	Living species.	Genera extinct.
Podocarpus, L'Hera																							22	+	+	+	+	+	+	+	+			
argillæ-Loedensis, Ett. & Gard																							+											
elatus, Gard																							+											
elegans (Phalacarpe) Ett. & Gard																							+											
coceusis, Ung.																							+											
Eyenasis, Crié.																							+											
gracilis, Sap																							+											
luceria, Ett. & Gard																							+											
lineuris, Sap																							+											
Lindleyana, Sap																							+											
proslma, Sap																							+											
Speolaevensis, Wat																							+											
Ginkgo, L. (Salisburia, Smlb																							+											
Bayeriana, Heyx																																		
buaervata, Lx																																		
coceusica, Ett. & Gard																																		
polymorpha, Lx																																		
Torreya, Arn																																		
dickaonioides, Dars																																		
Taxus, Ld																																		
Swansoni, Ett. & Gard																																		
Taxites, Bronga																																		
pectes, Bronga																																		
occidentalis, Newby																																		
Olriki, Heer																																		
Cephalotaxus, Sich. & Zuce.																																		
insignas, Heer																																		

a (60) Extratropical Southern Hemisphere; mountainous and Eastern Asia, mountains of tropical America (rare); absent from Southern Europe, Western Asia, Northern Africa, and North America.

b (1) China.

c (3-4) North America, Japan, China.

d (6-8) Temperate Northern Hemisphere.

e (4) Japan, China.

Table of distribution of Laramie, Senonian, and Eocene plants—Continued.

Table of distribution of Laramie, Senonian, and Eocene plants — Continued.

Table of distribution of Laramie, Senonian, and Eocene plants—Continued.

Species represented.

Thuya, L. (including Chamaecyparis, Spach.)—Continued.
   gracilis, Newby.
   interrupta, Newby.
Libocedrus, Endl. a.
   adpressa, Ett. & Gard.
Callitris, Vent. (including Widdring-
   tonia, Endl., Cupressinites, Bow.) b.
Brongniartii, Endl.
   complanata, Lx.
   curta (Bow.) Ett. & Gard.
   elongata, Bow. sp.
Ettingshauseni, Gard.
   globosa, Bow. sp
Reerii, Nap.
   recurvata, Bow. sp.
Reichii, Ett. sp.
   subfalciformis, Bow. sp.
Moriconia, Deb. & Ett.
   cyclotoxon, Deb. & Ett.
Inolepis, Heer.
   affinis, Heer.

a (8) 2 Chile, 2 New Zealand, 1 New Caledonia, 1 Japan, 1 China, and 1 California.

b (14) Africa, Madagascar, Australia, and New Caledonia.

Table of distribution of Laramie, Senonian, and Eocene plants—Continued.

Table of distribution of Laramie, Senonian, and Eocene plants.— Continued.

Species represented.	Laramie.	Senonian.	Eocene.	Summary of the foregoing.	Other formations in which found.

Column headers (Laramie): Bitter Creek, Golden, Raton Mountain, &c.; Carbon and Evanston; Port Union group.

Column headers (Senonian): Aix-la-Chapelle; Hartz District; Westphalia; Gosau formation, Austria; Lignites of Fuveau, Provence; Patoot, Greenland; Peace and Pine Rivers, British America; Vancouver and Orcas Islands; Paris Basin; Aix in Provence.

Column headers (Eocene): Arkansas de Brives; London clay; Monte Bolca, Pas-tello, and Promina; Other typical Eocene; Patoovine (Dewalquea, Sezanne, Mossons, &c.); Laramie; Senonian; Eocene.

Other formations: Lower (Cretaceous); Lower Cretaceous (the low Cretaceous); Cenomanian; Dakota group; Green River group; Oligocene; Miocene; Pliocene; Quaternary; Existing species; (Green extinct.)

Species list (left column):
Pandeae, Brongn.—Continued.
  dubius, Sap.
  glumaceus, Sap.
  nervosus, Sap.
  cladetus, Wat.
  ovatus, Sap.
  paucinervis, Wat.
  protogaeus, Wat.
  refertus, Sap.
  reticaceus, Sap.
  Roquei, Wat.
  Schimperi, Heer
  irrilevus, Sap.
Cyperaceae:
  Carex, L. a
  Bortlandi, Lx.
  Scirpus, L. b
    species, Dawa
  Cyperacites, Schp.
    Deleroisi, M.
    palaeostachyus, Sap.
    micranoides, Sap.
    Sezannensis, Sap.
  Rhizocaulon, Sap.
  Cyperorum, Sap.
  macrophyllum, Sap.
Ericaulonaceae:
  Eriocaulon, L. c
  purumun, Lx.

Footnotes:
a (500) Temperate and frigid regions, mountains in tropics.
b (300) All parts of the world.
c (100) All warmer parts of the world, except Northern Africa and Southern Europe; chiefly tropical.

B. A.

Table of distribution of Laramie, Senonian, and Eocene plants—Continued.

Table of distribution of Laramie, Senonian, and Eocene plants—Continued.

a (Africa 80) Europe, Northern Asia, tropical Africa, tropical Asia, North America, North America, Australia.
b (?) Temperate and tropical regions.
c (1) Tropical fresh waters, except Australia and Pacific Islands.
d (Typha) In Temperate and tropical regions.

Table of distribution of Laramie, Senonian, and Eocene plants—Continued.

Species represented.

Typhaelopum, Eng.—Continued.
  primaevum, Sap.
Sparganium, L.
  stritum, Sap.
  Stygianum, Heer.
Pandaneae:
  Pandanus, L. b
    Sinildus, Stohl.
  Landeripsis, Sap c.
    discrepta, Sap.
    g-odonatifida, Sap.
  Kaidacarpum, Carr.
    cretaceum, Heer.
Palmae:
  Latanites, M d
    parvulus, M.
  Calamopsis, Heer.
  Danua, Lx.
  Sabal Adans. (incl. Sabalites, Lx.)
    Andegavensis, Schp.
    Campbellii, Newby.
    fructifer, Lx. sp.
    Grayana, Lx. sp.
    Hastingiana (Ung.) Schp.
    imperialis, Daws.
    Latania, Rossm. sp.
    major, Ung.

a (76) Temperate and subtropical Northern Hemisphere and Australia.
b (30) Malay Archipelago, Mascarene and Seychelles Islands, Australia and Oceania (few); West Indies 1.
c (Carludovica 30) Tropical America, West Indies.
d (Latania 3) Mascarene Islands.
e (Calamus 200) Tropical and subtropical Asia, Africa (few), Australia (few), West Indies, Venezuela.
f (6) Tropical and subtropical North America, West Indies.

Table of distribution of Laramie, Senonian, and Eocene plants—Continued.

Table of distribution of Laramie, Senonian, and Eocene plants—Continued.

Species represented.
Geonomica, Lx.
Goldiana, Lx.
Schimperi, Lx.
Sonnbrakic, Lx.
Ungeri, Lx.
Oreodoxites, Lx. b.
plicatus, Lx.
Palmacites, Bronga
Aquensis, Sap.
manillatus, Bronga
mexanus, Wat
Axonresia, Wat
Antarthophyton, M
echinatus, Bronga
formosissa, M
Wetherellia, Bow
variabilis, Bow
Trirarpellites, Bow
scalaris, Bow
communis, Bow
crassus, Bow
cretus, Bow
gracilis, Bow
patens, Bow
Fugosus, Bow
Palaeocarpon
communc, Lx
compositum, Lx
curugatum, Lx
Mexicanum, Lx.

a (Grolonna 100) Tropical America.   b (Grolonna 5) Tropical America.

Table of distribution of Laramie, Senonian, and Eocene plants.—Continued.

Column headers (grouped):

- **Laramie:** Bitter Creek, Golden, Point of Rocks, Mountains, &c.; Carbon and Evanston; Fort Union group.
- **Senonian:** Aix-la-Chapelle; Harz District; Westphalia; (Gosau formation) Aus-tria; Lignites of Fuveau, Provence; Patoot, Greenland; Peru and Vine Rivers, British America; Vancouver and Queen Charlotte Islands.
- **Eocene:** Paris Basin; Aix in Provence; Arkoses de Brives; London clay; Sezanne flora, Pas-telle, and Gelinden; Monte Bolca; Other Typical Eocene; (Pasteur of Dunkirk), Sezanne, Gelinden (rich), &c.; Laramie; Senonian; Eocene.
- **Summary of the foregoing.**
- **Other formations in which found:** Lower than Lower Cretaceous; Lower Cretaceous; (now Comanchean); Cenomanian; Dakota group; Green River group; Oligocene; Miocene; Pliocene; (partly very); Living species; (Genera extinct).

Species represented.

Palmocarpon, Lx.—Continued.
subcylindricum, Lx.
truncatum, Lx.

Liliaceae:
Eolirion, Schenk
nervosum, Ros. & Mck.
Dracontes, Sap.
Brongnartii, Sap.
sepultus, Sap.
Majanthemophyllum, Web.?
athe-simium, M
cretaceum, Heer
laurvolatum, Heer
pusillum, Heer
Smilax, J.?
grandifolia, Ung.
Lyellii, Wat. sp
raiunifolia (Sap.) Seap.

Scitamineae:
Musophyllum, Göppd.
Axones-se, Wat.
longepaum, Sap.
speciosum, Sap.
Canuophyllites, Brongn.?
Ungeri, Wat.
Zingiberites, Heer?
dubius, Lx.

a (Dracæna 33) Warmer regions of the Old World.
b (Maianthemum 1) Temperate Northern Hemisphere.
c (16) Tropical and temperate regions.
d (Musa 29) Tropical regions of the Old World.
e (Canna 30) Tropical and subtropical America.
f (Zingiber 30) East Indies, Malay Archipelago, Mascarene and Pacific islands.

Table of distribution of Laramie, Senonian, and Eocene plants—Continued.

Species represented.	Laramie			Senonian						Eocene							Summary of the foregoing						Other formations in which found.									
	Bitter Creek, Golden, Raton Mountain, &c.	Carbon and Evanston.	Fort Union group.	Aix-la-Chapelle.	Harz District.	Westphalia.	Gosau formation, Austria.	Lignites of Provence, France.	Patoot, Greenland.	Peace and Pine Rivers, British America.	Vancouver and Orcas Islands.	Paris Basin.	Aix in Provence.	Arkansas de Rivers.	London clay.	Mounts Bolca, Tus-relio, and Promina.	Other typical floor ene.	Paleocene (Bernedoux, Sezanne, Suhzosa, &c.).	Laramie.	Senonian.	Eocene.	Lower than Creta.	Lower Cretaceous (below Cenomanian).	Cenomanian.	Dakota group.	Green River group.	Oligocene.	Miocene.	Pliocene.	Quaternary.	Living species.	Recent (extinct).

Anemophyllum, Ward?
  tenue, Ward
Hydrophiliaceae:
  lutifolia, Ward?
  Americana, Lx
  Paratropis, Sap.

Subclass II.—DICOTYLEDONS.

Division I.—Apetalæ.

Salicineæ:
Populus, Lc
  acerifolia, Newby.
  arctica, Heer.
  balsamoides Göpp.
  balsamoides extinua, Göpp
  cordata, Newby.
  cuneata, Newby
  decipiens, Lx
  denticulata, Heer
  glandulifera, Heer
  Heerii, Sap.
  Hookeri, Heer
  Lindgreni, Lx
  latior truncata, Al. Br.
  Ligeri, Sap.

a (Ammonum 50) Tropical Asia, Africa, Australia, and Pacific islands.
b (?) Tropical Asia, Japan, Australia, Mascarene Islands; tropical and subtropical Africa, Brazil.
c (10) Europe, middle and northern Asia, and mountains of tropical Asia, North America, Mexico.

Table of distribution of Laramie, Senonian, and Eocene plants—Continued.

Species represented.	Bitter Creek, Golden, Evanston Mountain, &c.	Fort Union group.	Aix-la-Chapelle.	Harz District.	Westphalia.	Gosau formation, Austria.	Lignites of Fuveau, Provence.	Patoot, Greenland.	Peace and Pine Rivers, British America.	Vancouver and Queen Charlotte Islands.	Paris Basin.	Aix in Provence.	Arkosee de Brives.	London clay.	Monte Bolca, Eas-telle, and Tournian.	Other typical Eocene.	Pliocene (Bracheux, Sézanne, Soissons, &c.).	Laramie.	Senonian.	Eocene.	Lower than Cret. Lower? (or in some for-mer (Cretaceous?) low Cretaceous.) (Common.) Dakota group. Green River group. Oligocene. Miocene. Pliocene. Quaternary. Living species. Genera extinct.
Populus, L.—Continued.																					
longior, Daws	+																	+	+		
melanaria, Heer																		+	+		
mutabilis oides, LX																		+			
tenuis ta, Wat. LX																		+			
neunoden, LX																		+			
mutabilis ovalis, Heer																			+		
mutabilis repanda-renata, Heer	+																		+		
Nebrascensis, Newby																		+			
nervosa, Newby																		+			
prinigenia, Sap																			+		
protozadarchii, Daws																			+		
rectiuervata, Daws																		+			
rhomboidea, LX																		+			
Richardsoni, Heer																			+		
retuudifolia, Newby																		+			
smilacifolia, Newby	+																	+			
Styxia, Heer																		+			
subrotundata, LX																		+			
Sueo-douensis, Wat.																			+		
tremulætormis, Ros. & Mek																			+		
trinervis, Daws																		+			
Ungeri, Lesq																			+		
Zaddachi, Heer																		+			
Populites M. eugcni																		+			
species, Daws																		+			
Salix, L. cuneata																		+			
ex clophylla, Heer																		+			
angusta? Al. Br																		+			

a (166) Temperate and frigid Northern Hemisphere; more rare in tropics; very few in Southern Hemisphere; 1 Chile; none in Malay Archipelago or South Pacific islands.

Table of distribution of Laramie, Senonian, and Eocene plants—Continued.

Salix, L.—Continued.
Aquensis, Sap
Axanitana, Ward
erehameria, Wat
Gortiniana, Heer
Harigii (Hank.) Schp
integra, Gopp
Islandira, Lx
longueqna, Sap
Pacifica, Daws
primaeva, Sap
Raeana, Heer
saeia, Sap
stupenda, Sap
tabellaris, Lx
Worthenii, Lx
Wv

Capulifera:
Fagus, L.o
dubia, Wat
roenqua, Wat
Frenalis, Unz
pseudoGiera, Daws
Wilkimannia, Ett
Castanea, Gaerta b
coceaica, Wat
Castanopsis (Lx.) Schp
Moorii (Lx.) Schp
mophihidendro (Gey.) Ett

a (12) Temperate and more frigid regions of both Hemispheres.
b (7) Temperate Northern Hemisphere, Asia, Europe, North America.
c (23) India, China, Malay Archipelago, 1 California

Table of distribution of Laramie, Senonian, and Eocene plants—Continued.

Table of distribution of Laramie, Senonian, and Eocene plants—Continued.

Species represented.	Bitter Creek, Golden, Raton Mountain, &c.	Carbon and Evanston.	Fort Union group.	Aix-la-Chapelle.	Harz District.	Westphalia.	Gosau formation, Austria.	Lignites of Provence, Provence.	Patoot, Greenland.	Peuce and Pine Rivers, Indian America.	Vancouver and Great Islands.	Paris Basin.	Aix in Provence.	Arkansas de Brives.	London clay.	Mount Dora, Pastelle, and Fremina.	Other typical Eocene.	Lahocene (Bruchsan, Sézanne, Soissons, &c.)	Laramie.	Senonian.	Eocene.	Lower than Critaceous.	Lower Cretaceous (below Cenomanian).	Cenomanian.	Dakota group.	Green River group.	Oligocene.	Miocene.	Pliocene.	Quaternary.	Living species.	Genera extinct.

Quercus—Continued.
fraxinifolia, Lx.
furcinervis (Rossm.) Ung.
Gedeli, Heer.
gracilis, Newby.
Haydenii, Heer.
interjeciola, Hos. & Mck.
Hookeri, Ett.
ilicitorme, Hos. & Mck.
Johnstani, Heer.
Lamberti, Wat.
Lanestani, Hos.
Langeana, Heer.
laurinica, Newby.
Legitosata, Hos.
Lonchitis, Ung.
longifolia, Hos.
Lyellii, Heer.
mastianta, Sap.
multinervis, Lx.
myrtillus, Heer.
negundoides, Lx.
nerifolia, Al. Br.
Olafseni, Heer.
paleophellos, Sap.
parallinervis, Wat.
parceserrata, Sap. & Mar.
Patooteusis, Heer.
pantcherrus, Wat.

Table of distribution of Laramie, Senonian, and Eocene plants—Continued.

Species represented:

Quercus—Continued.
paucinervis, Ros.
Peritii, Ett.
platania, Heer
platinervis, Lx.
perphillipsensis, Ett.
retracta, Lx.
rhamboidalis, Ros. & Mck.
salicina, Sap.
Salyorum, Sap.
spathulata, Wat.
spheroidoxli, Ros. & Mck.
stramineri, Lx.
Sullyi, Newby.
ternata, Sap.
euryphylla, Ros. & Mck.
Vaderiaich, Heer.
viburnifolia, Lx.
Victoriae, Daws.
Westfalica, Ros. & Mck.
Wilanii, Ros.
Dryophyllum, Deb.
Alnubeciatione, Deb.
Alberti-Magni, Deb.
Bottianum, Deb.
eampteroneurium, Deb.
crepatum, Lx.
Grppini, Deb.
credenum, Deb.
curtixelsen-o, (Wat.) Sap.
Dedunnieanum, Deb.

Table of distribution of Laramie, Senonian, and Eocene plants—Continued.

Table of distribution of Laramie, Senonian, and Eocene plants.—Continued.

Table of distribution of Laramie, Senonian, and Eocene plants.—Continued.

a (35.) Temperate and warmer regions of the world, except Australia.

Table of distribution of Laramie, Senonian, and Eocene plants—Continued.

Species represented.

Myrica, L. (Comptonia, Banks)—Continued.
Macbroniana, Sap.
Mc-inori, (Heer) Schp.
Merophioli, Ung.
parvula, Heer.
pedunculata, Wat.
platyphylla, Sap.
piaeox, Heer.
primaeva, Bos. & Mck.
pseudodryneia, Sap.
sabicina, Ung.
Saportana, Schp.
Schrakiana, Heer.
annua, Sap.
subdneiuriana, Sap.
subacias, Sap.
Saronineneis, Wat.
Torreyi, Lx.
Vinayi, Sap.

Juglandaceae:
Juglans, L.ϙ.
appreisa, Lx.
(?) elaeroa, Lx.
crassipes, Heer.
denticulata, Heer.
Harwoodeusis, Dawe.
Leconteana, Lx.

a (6) Temperate and subtropical Northern Hemisphere; 1 Europe and Middle Asia; 2 Eastern Asia and Japan; 4 or 5 North America, Canada, and California, to West Indies and Mexico.

Table of distribution of Laramie, Senonian, and Eocene plants—Continued.

Species represented:

Juglans, L.—Continued.
nigella, Heer
rhamnoides, Lx.
rugosa, Lx.
Saffordiana, Lx.
Schimperi, Lx.
Woodiana, Heer
Juglandites, St.
cornutus, Sap.
elegans, Gœpp.
obluselo-formis, Sap.
peramplus, Sap.
Carya, Nutt a
ambiquorum, Newby
Hecrii, Ett
Paleocarya, Sap
Martini, Sap.
Platanaceæ, l.
Platanus, l.
acerpodes, Gœpp
affinis, Lx.
Guillelmæ, Gœpp
Haydeni, Newby
latempliytha, Newby
Newberryana, Heer
nobilis, Newby
Raynoldsii, Newby

Table of distribution of Laramie, Senonian, and Eocene plants—Continued.

Species represented.

Platanus, L.—Continued.
  Raynoldsii heterifolia, Lx
  rhomboidea, Lx
Urticaceæ—
Artocarpus, Forst
  undulata, Hos
Artocarpidion, Sord
  conocephaloidea, Sap
  pomumformis, Sap
Artocarpidium, Ung
  Ephialtæ, Ett
  Stuarti, Ett
Ficus, L. G...
  angulata, Hos & Mck
  angustifolia, Hos
  nervis (Rossm.) Heer
  aurita, Heer
  arenacea, Lx
  autocarpoides, Lx
  asarifolia Lit
  atavina, Heer
  auriculata, Lx
  cinnamomoides, Lx
  crassinervis, Hos
  cretacea, Hos
  cunrata, Newby
  Dalmatica, Ett
  dracaenervis, Hos & Mck

a (10) Tropical Asia, via Pacific Islands, tropical Africa(?), Ceylon, Malay Archi..., and Pouronoma 26) Tropical South America.
b (Artocarpus cf. Conospora 18 and Pouronoma 26) Tropical South America.
c (600) Warmer regions of the globe. Most abundant in Malay Archipelago and Pacific islands. Few extra tropical Japan and Mediterranean region. Wanting in North America, except Mexico.

Table of distribution of Laramie, Senonian, and Eocene plants—Continued.

Species represented	Laramie			Senonian						Eocene						Summary of the foregoing			Other formations in which found									

Ficus, l.—Continued.
dentata, Hos.
elongata, Hos.
Göpperti, Heer.
gracilis, Hos.
Haydeni, Lx.
Hörneri, Hurr.
irregularis, Lx.
Jynx, Ung.
lanceolata, Hurr.
latifolia, Hos. & Mck.
longifolia, Hos.
maxima, Daws.
Morloti, Unz.
Murrayi, De la Harpe.
nervosa, Newby.
oblanceolata, Lx.
obscurata, Sap.
occidentalis, Lx.
planicostata, Lx.
planicostata 'goldiana, Lx.
planicostata latifolia, Lx.
platanifolia, Sap.
pseudopopulus, Lx.
pulcherrima, Sap.
Rertschii, Hos.
Schimperi, Lx.
Schlechtendali, Heer.
Smithsoniana, Lx.
spectabilis, Lx.
subtruncata, Lx.

Table of distribution of Laramie, Senonian, and Eocene plants—Continued.

Species represented	Laramie		Senonian									Eocene									Summary of the in region.	Other formations in which found.																

Ficus, L.—Continued.
tenuifolia, Heer.
tiliæfolia, Al. Br.
trunata, Heer.
trinerivia, Heer.
uncata, L.
uvaexta, Sap.
Verbo-kiana, Heer.
Proteoides, Sap.
crenulata, Sap.
insignis, Sap.
lacvæa, Sap.
Sezannensis, Sap.
Ficonium, Ett.
Planera, Gmel.
antiqua, Heer.
microphylla, Newby.
Ulmus, L.
antiquolina, Sap.
betulinoa, Sap.
Bouquarti, Peon.
dubia, Dawn.
Marionii, Sap.
neuroda, Wat.
oppositinervia, Wat.
plurinervia, Ung.
Santalaces:
Leptomeria, R. Br.

Bitter Creek, Golden, Raton Mountain, &c.
Carbon and Evanston.
Fort Union group.
Aix-la-Chapelle.
Harz District.
Westphalia.
(ossau formation, Aus-tria.
Lignites of Paveau, Provence.
Patout, Greenland.
Peace and Pine Rivers, British America, Vancouver and Orcas Islands.
Paris Basin.
Aix in Provence.
Alkoses de Brives.
London clay.
Alopgun Index, Ixabella, and Ptonimu.
Other typical Eocene.
Palawoene (Bracken, Sezanne, Sotzka,(&c.)
Laramie.
Senonian.
Eocene.
Lower than Creta-ceous.
Lower Cretaceous, the low (Wealdianum).
Neocomian.
Cenomanian.
Dakota group.
Green River group.
Oligocene.
Miocene.
Pliocene.
Quaternary.
Living species.
(recent) extinct.

a (i) Southern United States.
b (ii) Temperate Northern Hemisphere, mountains of tropical America.
c (iii) Asia.

Table of distribution of Laramie, Senonian, and Eocene plants—Continued.

Table of distribution of Laramie, Senonian, and Eocene plants—Continued.

Column groupings (headings rotated vertically):

Laramie: Bitter Creek, Golden, Eaton Mountain, &c.; Carbon and Evanston; Fort Union group.

Senonian: Aix-la-Chapelle; Harz District; Westphalia; Gosau formation, Austria; Lignites of Fuveau, Provence; Patoot, Greenland; Peeraruand Pine Rivers, British America; Vancouver and Orcas Islands.

Eocene: Paris Basin; Aix in Provence; Aikona de Hèves; London clay; Monnts Bolca, Fas-tello, and Promina; Other typical Eocene; Paleocene (Sézanne, Soissons, &c.); Laramie; Neurodan; Bovey.

Summary of the foregoing.

Other formations in which found: Lower than Crist.; Lower Cretaceous (the low Cenomanian); Cenomanian; Dakota group; Green River group; Oligocene; Miocene; Pliocene; Quaternary; Living species; (hence extinct).

Species represented.
Embothrites, Ung a.
Aquensis, Sap.
stenoptera, Sap.
Lomatia, R. Br b.
Bolerensis, Ung.
lation, Heer.
Lomatites, Sap f.
acerosus, Sap.
Aquensis, Sap.
Aquensis acuminatus, Sap.
Aquensis brevior, Sap.
Aquensis coriaceus, Sap.
Aquensis intermedius, Sap.
ninnatus, Sap.
Grevillea, R. Br. c
vera, Sap.
lipica, Sap.
myrtifolia, Sap.
nervosa, Heer.
provincialis, Sap.
rigida, Sap.
Persoonia, Sap d
Adenanthos, Labill e
species, Sap.
Kauzia, Heer.
Petrophiloides, Heer f
imbricatus, Heer.
Richardsoni, Heer.

a (Embothrium 4) South America, extratropical or Andes.
b (8) 3 Chile. 6 Australia.
c (650) Australia; 7 New Caledonia.
d (60) Australia. 1 New Zealand.
e (15) Extratropical Western Australia.
f (Petrophila 35) Australia.

Table of distribution of Laramie, Senonian, and Eocene plants—Continued.

Table of distribution of Laramie, Senonian, and Eocene plants—Continued.

Table of distribution of Laramie, Senonian, and Eocene plants—Continued.

Species represented.	Laramie			Senonian							Eocene									Summary of the foregoing.			Other formations in which found.									
	Bitter Creek, Golden, &c.	Fort Union group.	Carbon and Evanston; Union Mountain, &c.	Aix-la-Chapelle.	Harz District.	Westphalia.	Gosau formation, Aus-tria.	Lignites of Prov-ence.	Patoot (Greenland).	Patuxent (Pot. River), British America.	Vancouver and Orcas Islands.	Paris Basin.	Aix in Provence.	Arkansas deposits.	London clay.	Monte Bolca, Pan-tello, and Trinità.	Other (typical) Eocene.	Bournemouth (Ha-bury).	Sezanne; Kalaszin (Gel.)	Laramic.	Senonian.	Eocene.	Lower than Creta-ceous.	Lower Cretaceous (the low Cenomanian).	Cenomanian.	Dakota group.	Green River group.	Oligocene.	Miocene.	Pliocene.	Quaternary.	Living species.	Genera extinct.

Sassafras, Nerw.—Continued.
　primigenium, Sap.
　Schenki? Dawx.
　species, Dawx.
Ocoten, Aubl. (Oreodaphne, Nees) a
　apetiolia, Sap. & Mar.
Persea, Gærtn b.
　hantitidia, Lx.
　palæomorpha, Sap. & Mar.
　pedata, Lx.
　vetusta, Lx.
Cinnamomum, Blume c.
　Aquense, Sap.
　camphorefolium, Sap.
　ellipsoideum, Sap.
　emarginatum, Sap.
　grandifolium (Ett.) Schp.
　Heerii, Lx.
　lanceolatum (Ung.) Heer
　Leichardti, Ett.
　Mississippiense, Lx.
　affine, Lx.
　polymorphum, Al.Br.
　polymorphoides, McCoy
　Scheuchzeri, Heer
　Sezannense, Wat.
　Sextianum, Sap.
Daphnogene, Ung. emend.

a (200) Tropical and subtropical America; few in Canary Islands, South Africa, and Mascarene Islands.

b (100) Tropical and subtropical Asia, tropical and tropical America, Virginia, Chile; 1 Canary Islands.

c (30) Tropical and subtropical Asia, Japan, tropical Australia.

Table of distribution of Laramie, Senonian, and Eocene plants—Continued.

Species represented.	Laramie	Senonian	Eocene	Summary of the foregoing.	Other formations in which found.

Daphnogene, Ung.—cont'd—Continued.
Anglica, Heer
coriacea, Sap.
elegans, Wat.
longipes, Sap.
paradisea, Sap.
Kainensis, Sap.
Veronensis, M.

Monimineæ:
Monimiopsis, Sap a.
ulmernifolia, Sap.
amberstfolia, Sap.
fraterna, Sap.

Aristolochiaceæ:
Aristolochia, L.b.
cordifolia, Newby.

Polygonaceæ:
Coccoloba, L.c.
lævigata, Lx.

Nyctagineæ:
Pisonia, L.d.
eocenica, Ett.
racemosa, Lx.

Division II.—Polypetalæ.

Corniaceæ:
Nyssa, L.e.
lanceolata, Lx.

a (Monimia ?) Mascarene islands.
b (89) Tropical America; few in Asia, Pacific and Mascareno Islands.
c (88) Warmer and temperate regions of the whole globe.
d (90) Tropical America; chiefly tropical: few in Mexico and Florida.
e (86) Temperate and warm Eastern North America, Eastern Himalayas, Malay Archipelago.

Table of distribution of Laramie, Senonian, and Eocene plants—Continued.

	Laramie.			Senonian.							Eocene.									Summary of the foregoing.			Other formations in which found.											

Species represented.

Cornus, Lu ....
Nelumbium, Heer ....
Nectandra?, Schp ....
platyphylla, Sap ....
rhamnifolia, W'rh ....
Stauloef, Heer ....
microclidens, Ls ....
Thalesia, Heer ....

Araliaceæ:
Hedera, Lu ....
cuneata, Heer ....
Philiberti, Sap ....
prisca, Sap ....
Cussonia, Thunb ....
rediviva, Sap ....
recinervia, Sap ....
Panax, Lf ....
globulifera, Heer ....
macrocarpa, Heer ....
Aralia, Le ....
acerifolia, Ls ....
argutidens, Sap ....
bicornis, Sap ....
calyptrocarpa, Sap ....
cordifolia, Sap ....
crenata, Sap ....

a (25) Europe, Asia, temperate America; few in Mexico; 1 Peru.
b (2) temperate and subtropical Northern Hemisphere; from Canaries to Japan; 1 Australia.
c (11) Tropical and Southern Africa, Mascarene Islands.

d (25) Tropical and Eastern Asia to Manchooria, tropical Africa, Pacific Islands, New Zealand, Australia.
e (30) Tropical and Eastern temperate Asia, North America, Mexico, Japan, Malay Archipelago.

Table of distribution of Laramie, Senonian, and Eocene plants—Continued.

	Laramie.					Senonian.							Eocene.										Summary of the foregoing.		Other formations in which found.												
Species represented.	Bitter Creek, Golden, Raton Mountain, &c.	Carbon and Evanston.	Fort Union group.	Aix-la-Chapelle.	Hart District.	Westphalia.	Gosau formation, Austria.	Lignites of Fuveau, Provence.	Patoot, Greenland.	Peace and The Rivers, British America.	Vancouver and Texas lignites.	Paris Basin.	Aix in Provence.	Arkansas de Rives.	London clay.	Monte Bolca, Tas-tello, and Promina.	Other (1) plant Eocene.	Theoceae (Blackburn), Sezanne, Suissonn.(fol).	Laramie.	Senonian.	Eocene.	Lower than Creta-ceous.	Lower Cretaceous (the-low Cenomanian.)	Cenomanian.	Dakota group.	Green River group.	Oligocene.	Miocene.	Pliocene.	Quaternary.	Living species.	(towers extinct).
Araliaceae, L.—Continued.																																
dentriata, Hos. & Mck																			+		+											
denticulata, Hos. & Mck																					+											
hederacea, Sap																			+		+											
Lessigiana, Sap. & Mar																					+											
microphyllla, Hos. & Mck																					+											
multifida, Sap																					+											
notata, Lx																			+		+							+				
primigenia (Dr. la Harpe) Heer											+										+											
pungens, Lx																			+		+											
racemifera, Sap																					+											
robusta, Sap																					+											
Sezannensis, Sap																		+			+											
spiculosa, Sap																					+											
triloba, Newby			+			+		+											+		+											
tripartita, Lx						+															+											
vernalosa, Sap																		+			+											
Wuigattensis, Heer																					+								+			
Onagrarieæ:																																
Trapa, L. c			1																									+	+			
microphylla, Lx			1																													
borealis, Heer			B. A																									+				
Melastomaceæ:																																
Melastomites, Ung. b																																
americanus, Hos. & Mck						+															+											
Myrtaceæ:																																
Eugenia, L. c																																
Apollinis, Ung																													+			
Heliæ, Ung																													+			

a (3-4) Middle and Eastern Europe, tropical and subtropical Asia and Africa.   b (Melastoma. 40) Tropical Asia, Northern Australia, Oceanica, 1 Seychelles.   c (700) Tropical and subtropical America, tropical Asia; few Australia and Africa.

Table of distribution of Laramie, Senonian, and Eocene plants.—Continued.

a (100) Extratropical Western South America; tropical America (fewer); 8 Australia: 1 or 2 New Caledonia.
b (18) Pacific Islands (New Zealand, New Zealand—Hawaii); 1 tropical Australia. 1 Indian Archipelago, 1 South Africa.
c (100) Australia, Indian Archipelago (few).
d (Callistemon (2) Australia; 1 or 2 New Caledonia.
e (Leptospermum 25) Australia; New Zealand; New Caledonia, Indian Archipelago (few).
f (80–90) Tropical Eastern Hemisphere, few tropical America.

Table of distribution of Laramie, Senonian, and Eocene plants—Continued.

Column headings (read vertically):

Species represented:

Terminalia, L.—Continued.

Hamamelideæ:
Liquidambar, L.?
Hamamelites, Sap?

Saxifrageæ:
Ribes, L.?
Cercidiphyllum, Sap.

Rosaceæ:
Amelanchier, Lindl?
Cotoneaster, Medik?
Crataegus, L.?

Laramie:
Bitter Creek, Golden, Raton Mountain, &c.
Carbon and Evanston.
Fort Union group.

Senonian:
Aix-la-Chapelle.
Harz District.
Westphalia.
Gosan formation, Austria.
Lignites of Paruu, Provence.
Patoot, Greenland.
Peace and Pine Rivers, British America.
Vancouver and Orcas Islands.

Eocene:
Paris Basin.
Aix in Provence.
Arkose de Bitven.
London clay.
Monnia Bolca, Pastello, and Promina.
Other typical Eocene.
Pahoerac (Brackiouz, Sezaune, Solsoous, Gel.)
Laramie.
Senonian.
Eocene.

Summary of the foregoing.

Other formations in which found:
Lower than Cretaceous.
Lower Cretaceous, low Cenomanian.
Cenomanian.
Dakota group.
Green River group.
Oligocene.
Miocene.
Pliocene.
Quaternary.
Living species.
(now extinct)

Footnotes:

a (2) tropical to temperate North America, 1 Asia Minor.
b (1) Hamameris, 2) Eastern North America, 1 Japan.
c (56) Temperate Europe, Asia and America; America.
d (2) Temperate Eastern Australia.
e (4) Asia Minor, Japan, North America.
f (15) Europe, Northern Africa, Middle and Western Asia, Siberia, mountains of the East Indies, Mexico. Hamamelis, North America.
g (56) Europe, Western and Northern Asia, Japan, North America (Canada to Mexico); Andes, New Granada.

Table of distribution of Laramie, Senonian, and Eocene plants—Continued.

Species represented.	Laramie.						Senonian.						Eocene.						Summary of the foregoing.								Other formations in which found.								
	Bitter Creek, Golden, Raton Mountain, &c.	Carbon and Evanston.	Fort Union group.	Aix-la-Chapelle.	Harz District.	Westphalia.	Gosau formation, Austria.	Lignites of Puteaux, Provence.	Patoot, Greenland.	Peace and Pine Rivers, British America.	Vancouver and Orcas Islands.	Paris Basin.	Aix in Provence.	Arkoses de Bitves.	London clay.	Monte Bolca, Pas tello, and Promina.	Other typical Eocene.	Paleocene (Brabbux), Sezanne, Noissons (?, tel.)	Laramic.	Senonian.	Eocene.	Lower than Cretaceous.	Lower Cretaceous (below Cenomanian).	Cenomanian.	Cretaceous.	Dakota group.	Green River group.	Oligocene.	Miocene.	Pliocene.	Quaternary.	Living species.	(Genera extinct).		
Crotozon, L.—Continued.																																			
fragranoides, Heer																																			
nobilis, Sap																																			
Leguminosites:																																			
Acacia, Wild a.																																			
ambigua, Sap																																			
Aquensis, Sap																																			
lingulae, Sap																																			
Coracurnea, Sap																																			
midateroides, Sap																																			
lacerata, Sap																																			
longipinna, Sap																																			
oleasperma, Sap																																			
palurnondes, Sap																																			
acuminea, Sap																																			
Mimosa, L b																																			
Cercis, L c																																			
antiqua, Sap																																			
borealis, Newby																																			
truncata, Lx																																			
Podogonium, Lx																																			
Americanum, Lx																																			
Cassia, L d																																			
ambigua, Ung																																			
australis, Heer																																			
corthna, Heer																																			
Cookii, Kit																																			

a (420) Africa (chiefly), Australia, and other warm regions.
b (220) America (warmer regions), Africa and Asia (f-w).
c (3-4) Europe, temperate Asia, Japan, North America.
d (400) Warm regions Bogoria and Chile to Mexico, Eastern United States, Africa (all parts), tropical Asia, Australia (not in New Zealand nor Tasmania).

Table of distribution of Laramie, Senonian, and Eocene plants—Continued.

Cassia, L.—Continued.

Species represented.

Table of distribution of Laramie, Senonian, and Eocene plants—Continued.

Species represented:

- Dalbergia, Ung. f.—Continued.
- pumilio, Heer
- Micropodium, Sap.
- affine, Sap.
- oligospermum, Sap.
- Phaseolites, Ung a.
- clitoriaeformis, Sap.
- obtusiusca, Sap.
- orbicularis, Ung
- Ervites, Sap b.
- primaevus, Sap.
- Colutea, L c.
- parvifoliata, Sap.
- prolagoea, Heer
- Leguminosites, Brongn.
- arachioides, Lx.
- cassioid-s, Lx.
- dentatus, Heer
- frigidus, Heer
- gastrolobianus, Sap.
- Kennedya, Ett
- orbicularis, Heer.
- Proteoides, Heer.
- Sprengeli, Heer
- species (fructus, etc.)
- Faboidea, Bow
- 25 species
- Anacardieae
- Pistacia, L d.

Column groups (rotated headers):

Laramie:
- Bitter Creek, Golden, Iron Mountain, &c.
- Carbon and Evanston.
- Fort Union group.

Senonian:
- Aix-la-Chapelle.
- Harz District.
- Westphalia.
- Gosau formation, Aus- tria.
- Lignites of Fuveau, Provence.
- Patoot, Greenland.
- Peace and Pine Rivers, British America.
- Vancouver and Orcas Islands.
- Paris Basin.

Eocene:
- Aix in Provence.
- Arkoses de Brives.
- London clay.
- Monte Bolca, Pas- tello, and Promina.
- Other typical Eocene.
- Paleocene (Sezanne, Soissons, Gel.)
- Laramie.
- Senonian.
- Eocene.

Summary of the foregoing:
- Lower than Cret a.

Other formations in which found:
- Lower Cretaceous (be- low Cenomanian).
- Cenomanian.
- Dakota group.
- Green River group.
- Oligocene.
- Miocene.
- Pliocene.
- Quaternary.
- Living species.
- (recent extinct)

6 GEOL——32

a (Phaseolus 50) Warmer regions (including temperate North America).
b (Vicia 186) Temperate Northern Hemisphere and South America.
c (7–8) Middle and Southern Europe, temperate and subtropical Asia.
d (6) Mediterranean region, Canary Islands, warmer Western Asia, Mexico.

FLORA OF THE LARAMIE GROUP.

Table of distribution of Laramie, Senonian, and Eocene plants—Continued.

Table of distribution of Laramie, Senonian, and Eocene plants—(Continued.)

Species represented.	Laramie.			Senonian.					Eocene.									Summary of the foregoing.				Other formations in which found.											

Acer L.—Continued.
 sclentatum, Heer
 gracilescens, Lx
 Sextianum, Sap
 tribulatum productum, Lx
Sapindus, Lx
 semulus, Heer
 anceps, Heer

 affinis, Newby
Cupaninidea, Bow b
 caudatus, Lx
 dysgenophyllus, Sap
 membranaceus, Newby
 Morisoni, Heer
 obtusifolius, Lx
 undulatus, Mr.
 Velhersensis, Sap
 corrugatus, Bow
 depressus, Bow
 grandis, Bow
 inflatus, Bow
 lobatus, Bow
 pygmaeus, Bow
 subangulatus, Bow
 tumidus, Bow
 antiqua, Daws
Ewalius, L c

a (40) Tropical regions of both hemispheres; subtropical, more rare.
b (Onpania 30) Tropical regions of both hemispheres; rared in Africa.
c (14) North America, Mexico, mountains of New Granada, Himalayas, Persia, Malay Peninsula.

Table of distribution of Laramie, Senonian, and Eocene plants—Continued.

Species represented.	Laramie.					Senonian.								Eocene.													Summary of the foregoing.			Other formations in which found.									
	Bitter Creek, Golden, Raton Mountain, &c.	Carbon and Evanston.	Fort Union group.	Aix-la-Chapelle.	Harz District.	Westphalia.	Gosau formation, Austria.	Lignites of Patoot, Province.	Patoot (Greenland).	Peace and Pine Rivers, British America.	Vancouver and Orcas Islands.	Paris Basin.	Aix in Provence.	Arkoses de Brives.	London clay.	Monntia fulva, Ina-tella, and Pruming.	Other typical Eocene.	Sezanne, Nebraska (Gol.).	Fort-horeux, Sezanne.	Laramie.	Senonian.	Eocene.	Lower than Creta-ceous.	Lower (?) Cretaceous (low Cenomanian).	Cenomanian.	Dakota group.	Green River group.	Oligocene.	Miocene.	Pliocene.	Quaternary.	Living species.	Genera extinct.
Ampelideæ:																																	
Vitis, L. (including Cissus, L.) a	5	1																		5								+	+		+	+	
ampelopsoides, Sap																																	
lævigata, Lx		+																												+			
lobato-crenata, Lx		+																															
Olriki, Heer	+	+																		+													
primæva, Sap																						+											
Sezanneuse, Sap																					+	+											
sparsa, Lx																						+											
(rotundata, Heer																																	
Cissites, Heer																												+	+				
Inœrus, Sap																												+					
Chondrophyllum, Bruu																												+	+				
hederæforme, Heer																												+ +					
Rhamneæ:																																	
Pomaderrieæ, Ett b																																	
Bankeii, Ett																																	
Gœnoplisia, Lx	3								+											3													
cretacea, Dawe																																	
podromus, Heer																																	
species? Hos. & Mck																																	
Rhamnus, L d	10	3	3			+				+	+									10									+	+		+	+
alaternoides, Heer																+																	
arctoidea, Sap																																	
Cleburni, Lx																																	
deformatus, Lx		+																															
discolor, Lx		+																															
Goldianus, Lx		+																															

a (220) Tropical, subtropical, and temperate regions; rarest in America.
b (Pomaderrie, 16) Southern and Eastern Australia; New Zealand.
c (28) Tropical and temperate North America, especially westward.
d (60) Warm and temperate Europe, Asia, and America (rarer in tropical); tropical Africa, Australia, Pacific islands.

Table of distribution of Laramie, Senonian, and Eocene plants—Continued.

a [10] Northern India, Java, China, eastern tropical Africa, tropical and temperate North America.

b [50] Tropical Asia and America, Africa (rare), Australia (very rare); warm, extratropical regions of both Hemispheres.

Species represented:

Rhamnus, L.—Continued.
marginalis, Lx.
oboratus, Lx.
parvifolius, Newby.
Pfaffianus, Heer.
rectinervis, Heer.
Rossheri, Kit.
Rossmässleri, Ung.
salicifolius, Lx.
species, Heer & Nick.
species, Daws.
Rhamnites, Newby.
coccineus, Newby.
Perchonia, Merk.
multinervis, Al. Br.
Zizyphus, Juss. b
Bockwitzii, Lx.
distortus, Lx.
fibrillosus, Lx.
Groenlandicus, Heer.
hyperboreus, Heer.
interpiolinus, Heer.
Meekii.
Meigsii (Lx.), Ship.
paradisiacus, Heer.
pseudolotus Ungeri, Sap.

Table of distribution of Laramic, Senonian, and Eocene plants—Continued.

Species represented:

Zizyphus, *Juss.*—Continued
  Reducentrii, Sap
  remotidens, Sap
  Cuperi, Heer
  velutina, Heer
Paliurus, *Juss a*
  affinis, Heer
  Colombi, Heer
  tenuifolius, Heer
  zizyphoides, Lx
Celastrinae:
Celastrus, Lb
  Celastrus, l b
  Andromedæ, Ung
  arctica, Heer
  læuiis(?), Heer
  lanulæ-formis, Sap
  scrophulatus, Ung
  Fälger-thensis, Ett
  pseudo-Bruckmanni, Sap
  Celastrinites, Sap
  arctocarpides, Lx
  fallax, Sap
  Hartogianus, Sap
  lævigatus, Lx
  legitimus, Sap
  venulosus, Sap
Celastrophyllum, Ett
  ervnatum, Heer
  Cunninghami, Ett
  Boseleydi, Sap
  lanceolatum, Heer

*a (2) 1 Southern Europe and Western Asia; 1 Southern China.
b (19) Mountains of India, China, Japan; few in North America and Australia; 1 Madagascar.*

Table of distribution of Laramic, Senonian, and Eocene plants.—Continued.

Species represented.

Celastrophyllum, Ett.—Continued
  repandum, Sap.
  reticulatum, Sap.
  acutidens, Sap.

Ilicineæ:
  Ilex, L. a
    inordita, Heer
    dissimilis, Lx.
    microphylla, Newby.
    Patootensis, Heer
    quercina, Sap.
    Sotzeana, Sap.
    stenophylla, Ung.

Simarubeæ:
  Ailanthus, Desf. b
    lanceoa, Sap.
    montissima, Sap.
    priaca, Sap.
    tremula, Sap.

Zygophylleæ:
  Guajacum, M c
    eueryca, M.
    Heerii, M.

Malpighiaceæ:
  Malpighiastrum Ung. d
    Indusiaeum, Ett.

Tiliaceæ:
  Apeibopsis, Heer e

a (165) Temperate and tropical regions, chiefly South American: rare in Africa and Australia.
b (3) East Indies and China.
c (Guaiacum 8) Tropical and subtropical North America.
d (Malpighia 20) Tropical America.
e (Apeiba 5) Tropical America.

Table of distribution of Laramie, Senonian, and Eocene plants—(Continued.)

Table of distribution of Laramie, Senonian, and Eocene plants.—Continued.

Species represented.	Laramie.			Senonian.									Eocene.									Summary of the foregoing.			Other formations in which found.													

Dombeyopsis, Ung.—Continued.
  anisoloba, M.
  lilioides, M.
  trivialis, Lx
  Hereopernice, Heer *a*
  impegualis, Sap.
Steculia, *l b*
  labrusca, (Ung.
  Michelui (Wat.) Sap.
  modesta, Sap.
  tenuiloba, Sap.
  variabilis, Sap.
Fraenstoria, M.
  angaria, M.
  celtifformis, M.
  Gilcalius, M.
  clavæfonnia, M.
  cucurbitina, M.
  gastrioides, M.
  gigantea, M.
  Lagenaria, M.
  Megaptrye, M
  Melo, M
  poiniformis, M.
  pyramidalia, M.
  pyriformis, M.
  rotunda, M.
  Zapoaua, M
Credneriaceæ
  Credueria, Zenk.

Table of distribution of Laramie, Senonian, and Eocene plants—Continued.

Species represented.	Laramie.			Senonian.										Eocene.								Summary of the foregoing.			Other formations in which found.											
	Bitter Creek, Golden, Raton Mountain, &c.	Carbon and Evanston.	Fort Union group.	Aix-la-Chapelle.	Harz Distelet.	Westphalia.	Gosau formation, Austria.	Lignites of Puvean, Provence.	Patoot, Greenland.	Peruvian Pine River, British America.	Vancouver and Texas Islands.	Paris Basin.	Aix in Provence.	Argiles de Hérya.	London clay.	Mount Index, Tasmania, &c.	Other typical Eocene.	Paleocene (Bracheux, Sézanne, Soissons, &c.)	Laramie.	Senonian.	Eocene.	Lower than Creta.	Lower (cretaceous the low Cenomanian).	Cenomanian.	Dakota group.	Green River group.	Oligocene.	Miocene.	Pliocene.	Quaternary.	Living species.	Genera extinct.
Credneria, Zenk.—Continued.																																
acuminata, Hampe.																																
denticulata, Zenk.																																
integerrima, Zenk.																																
oblonga, Schp.																																
subtriloba, Zenk.																																
tenuinervis, Ett.																																
triacuminata, Hampe.																																
Wendtiæa, Hos.																																
Protophyllum, Lx.																																
boreale, Dawx.																																
Leconteanum, Lx.																																
Knowltoni, Dawx.																																
rugosum, Lx.																																
Anisophyllum, Lx.																																
species, Iowa.																																
Malvaceæ:																																
Bombax, Lx.																																
Hreii, Ett.																																
Michellii, Ett.																																
septiflorum, Sap.																																
Ternstroemiaceæ:																																
Samyda, Wildb.																																
robusta, Sap.																																
Pittosporeæ:																																
Pittosporum, Banks c																																
Paraskii, Ett.																																
latifolium, Sap.																																
laurinum, Sap.																																

a (10) Tropical America; Tropical Africa; Tropical Asia.
b (60) Tropical and subtropical Asia and America.
c (50) Africa, warmer parts of Asia; Pacific Islands, Australia, New Zealand.

Table of distribution of Laramie, Senonian, and Eocene plants — (Continued.)

Species represented.

Pittosporum, Banks—Continued.
priscum, Ett.
pulchrum, Sap.
pandurum.
Nymphaeaceae:
Nelumbium, Juss a.
Buchii, Ett.
Lakoelii, Lx.
nymphaeoides, Ett.
resnidium, Lx.
Nymphaea, Lb.
Charpentieri, Heer.
gypsorum, Sap.
parvula, Sap.
Nymphaeites, St.
arethusae (Brongn.) St.
Peltophyllum, M.
nelumbioides, M.
Menispermaceae:
Menispermites, Lx c.
reniformis, Daws.
Cocculus, DC d.
Dumontii, Sap.
Knii, Sap.
Macclintockii, Heer.
cretacea, Heer.
Anonaceae:
Anona, L e.
Altenburgensis, Ung.

a (29) Tropical Asia, tropical Australia.
b (29) Tropics and Northern Hemisphere, few South Africa and Australia.
c (Menispermum 2) 1 North America, 1 temperate Eastern Asia.

d (18) Tropical Africa and Asia; China; 2 warmer parts of North America.
e (50) Tropical America; 2-3 tropical Africa and Asia.

Table of distribution of Laramie, Senonian, and Eocene plants — Continued.

Species represented.

Anona, L.—Continued.
 lignitum, Ung ......
 robusta, Lx ......
Asimina, Adans.d ....
 eocenica, Lx ....
 leiocarpa, Lx ....
Magnoliaceæ:
 Liriodendron. L. b
 Meekii, Heer ......
Magnolia, L. f ......
 angustifolia, Newby .....
 attenuata, Web ......
 Brownii, Ett ....
 cordifolia, Lx ....
 Inglefieldia, Lx ....
 inæqualis, Sap. ......
 insurfolia, Lx ......
 Lesleyana, Lx ......
 Ligerina, Sap. ......
 magnifica, Rawa ......
 ovalis, Lx ......
 rotundifolia, Newby ....
 tenuinervis, Lx ....
 tenuifolia, Lx ....
 Tourreai, Ett ....
Ranunculaceæ:
 Dewalquea, Sap. & Mar ......
 Gelindensis, Sap. & Mar ....
 Groenlandica, Heer ....
 Haldemann, Sap. & Mar ......

a (-e) North America (including Mexico), Central America (Cuba). b (1) North America. c (14) North America, Mexico, Japan, China, Himalayas.

Table of distribution of Laramie, Senonian, and Eocene plants—Continued.

Species represented.

Dewalquea, Sap. & Mar.—Continued.
　Halle-aiana angustifolia, Hos. &
　　Mek.
　Halle-aiana inaifolia, Hos. & Mek.
　insignis, Hos. & Mek.

Division III.—Gamopetalae.

Bignoniaceae:
　Catalpa, Jaeg a.
　　rixaefolia, Newby.
　　macrosperma, Sap.
Solanaceae:
　Solanites, Sap.
　　Brongniartii, Sap.
Convolvulaceae:
　Porana, Barnb.
　　Boleraea (Ung.) Schp.
　　po-ellonides (Al.) Schp.
Asclepiadaceae:
　Gomphocarpus, R. Br. (including Are-
　　rates, Ell.) c
　　arctica, Heer.
Apocynaceae:
　Echitonium, Ung d
　　Scanurnae, Wat.
　Apocynophyllum, Ung c.
　　camulum, Hos. & Mek.

a (6) China, Japan, North America, West Indies.
b (6) East Indies, Malay Archipelago, Australia.
c (80) Southern and tropical Africa, Arabia; North and Central America.
d (Echites 25) Tropical America.
e (Apocynum 5) Southern Europe, temperate Asia, North America.

Table of distribution of Laramie, Senonian, and Eocene plants—Continued.

Species represented.	Laramie.			Senonian.									Eocene.												Summary of the foregoing.			Other formations in which found.								
	Bitter Creek, Hollow, Raton Mountain, &c.	Carbon and Evanston.	Fort Union group.	Aix-la-Chapelle.	Haiz District.	Westphalia.	Gosau formation, Aus. tria.	Lignites of Patagonia, Provence.	Patoot, Greenland.	Peace and Pine Rivers, British America.	Vancouver and Queen Charlotte Islands.	Paris Basin.	Aix in Provence.	Aix-les-de-Rivers.	London clay.	Monte Bolca, Tur. telic, and Provins.	Other typical Eocene.	Paleocene (Throuhoux, Sezanne, Sézanne, Sezanne, &c.)	Laramie.	Senonian.	Eocene.	Lower than Creta.	Lower Cretaceous (Urgonian).	low Urgonian.	Cenomanian.	Dakota group.	Green River group.	Oligocene.	Miocene.	Pliocene.	Quaternary.	Living species.	(extinct exhibit)

Apectophyllum, Ung.—Continued
  Etheridgii, Ett.
  nerifolium, Heer
  plumerti-folium, Ett.
  subrepandum, Marck
  Sumatrense, Heer

Nerium, La
  Parisiense, Sap
  repertum, Sap
  Rohlii, Marck

Tabernæmontana, L. b.
  primigenia, Ett

Oleaceæ:
  Olea, L. c.
    proxima, Sap
  Notelea, Vent f.
  Fraxinus, L. c.
    dentitata, Heer
    curvata, Lx
    exilis, Sap
    præcox, Heer

Styraceæ:
  Symplocos, L. f
  Bursinaum, Sap

Ebenaceæ:
  Diospyros, L. g

a (7–2) Mediterranean region, subtropical Asia, Japan.
b (119) Tropical regions of the world.
c (78) Tropical and Middle Asia, Mediterranean region, tropical and South Africa, Macaronian islands, New Zealand.
d (8) Australia.
e (50) Temperate and subtropical Northern Hemisphere.
f (66) Warmer regions of Asia, Australia, and America (none in Africa).
g (153) Tropical regions of the world; temperate Asia and America.

Table of distribution of Laramie, Senonian, and Eocene plants — Continued.

Column groups (left to right): **Laramie** (Bitter Creek, Golden; Point of Rocks Mountain, &c.; Carbon and Evanston; Fort Union group) — **Senonian** (Aix-la-Chapelle; Harz District; Westphalia; Gosau formation, Austria; Lignites of Fuveau, Provence; Patoot, Greenland; Peace and Pine Rivers, British America; Vancouver and Queen Charlotte Islands; Paris Basin) — **Eocene** (Aix in Provence; Alkossic Diatoms; London clay; Menatis Dolen, Tune, Gelinden, and Tronum; Other typical Eocene; Paleocene (Bracheux, Sezanne, Soissons, Gel.); Laramie; Senonian; Eocene) — **Summary of the foregoing** — **Other formations in which found** (Lower than Cretaceous; Lower Cretaceous (the low Cenomanian); Cenomanian; Dakota group; Green River group; Oligocene; Miocene; Pliocene; Quaternary; Living species; (Desert extinct))

a (20) Tropical and North America.

Species represented:

Diospyros, L.—Continued.
  mkscripta, Sap.
  ambigua, Sap.
  brachysepala, Al. Br.
  rotundata, Sap.
  discreta, Lsx.
  ficoidea, Lsx.
  Harpari, Heer.
  involucrae, Sap.
  utilis, Dawa.
  oocarya, Sap.
  gracuneor, Sap.
  primava, Heer.
  rhododendrifolia, Sap.
  virgosa, Sap.
  Steenstrupi, Heer.
  Vancouverensis, Daws.
  vetusta, Heer.
  Wodani, Ung.
  species, Daws.
Sapotaceae—
  Bumelia, Sw a
    oblongifolia, Ett.
    Dreamnn, Unz.
    ellipathiana, Sap.
  Sapotacites, Ett.
    ambignus, Ett.
    crassipes, Heer.
    and, Sap.
    hyperboreus, Heer.

Table of distribution of Laramie, Senonian, and Eocene plants—Continued.

Species represented.	Laramie			Senonian										Eocene									Summary of the foregoing.		Other formations in which found.												

Sapotaceæ, Ett.—Continued.
 nervillosus, Heer
 reticulatus, Heer
 retusus, Heer
 sideroxyloides, Ett.
 vaccinioides, Ett.
Myrsineæ, l. c.
 Myrsine, l. c.
 acuminata, Sap.
 confusa, Sap.
 formosa, Heer
 liocarpa, Sap.
 marginata, Sap.
 præstiperata (Ung.) Sap.
 spinulosa, Sap.
 Vinoyana, Sap.
Ericaceæ:
 Rhododendron, L. b
 Saturni, Ett.
 Andromeda, L. c
 afavia, Sap.
 Grayana, Heer
 macrosmata, Sap.
 protogæa (Ung.) Sap.
 abbreviata, Sap.
 Leucothoe, Don d
 arctuervix, Sap.

a (60) Tropical Asia, Africa, and America; a few in Japan, extratropical Africa, Australia, Atlantic islands, and New Zealand.

b (130) Mountains of Europe, Asia, Malay Archipelago, and North America. (Most abundant in the Himalayas.)

c (1) Temperate and subarctic regions of the Northern Hemisphere. (Genus generally said to include Lyonia, Xenobia, and Pieris, ranging to Mexico, West Indies and Malay peninsula.)

d (9) Eastern North America and Japan.

Table of distribution of Laramie, Senonian, and Eocene plants.—Continued.

Species represented:

Leucothoe, Hem.—Continued.
  linaria, Sap.
  pulchra, Sap.
  subterranea, Sap.
Gaudichi, Le
  excelsica, Ett.
Vacciniaceæ:
  Vaccinium, L. b
  Arbutifolium, Ung.
  Asperue, Sap.
  obscurum, Sap.
  parvulum, Sap.
  proximum, Sap.
  reticulatum, Al. Br
  acerenodum, Sap.
Compositæ:
  Carduuites, Sap c
  prisca, Sap
  Cynarites, Heer
  Cypserena, Sap.
Valerianeæ:
  Valerianellites, Sap d
  capitatus, Sap.
Rubiaceæ:
  Gardenia, L c
  Gardneri, Heer

a (90) North America, Andes, mountains of India, and Malay Archipelago; few in Australia and New Zealand; 1 Japan.
b (100) Temperate Northern Hemisphere; mountains of the tropics. (None in south temperate zone.)
c (Parthenium, 6). North and Central America, and West Indies. (Reference very doubtful.)
d (Valerianella, 47). Europe, Northern Africa, Western Asia, and North America.
e (Mediterranean to Mediterranean region.)
e (60) Tropical and subtropical regions of the Old World; Pacific Islands.

Table of distribution of Laramie, Senonian, and Eocene plants—Continued.

Species represented.

Caprifoliaceæ:
Viburnum, Lx
asperum, Newby.
aticuatum, Heer
Dakotense, Lx
Dentoni, Lx
dichotomum, Lx
giganteum, Sap.
Goldianum, Lx
Lakesii, Lx
lanceolatum, Newby.
marginatum, Newby
multinerve, Heer
Nordenskiöldi, Heer
platanoides, Lx
pubescens, Pursh
rotundifolium, Lx
subintegrum, Hos. & Mek
ellifolium, Sap. & Mar.
Whympcri, Heer
zizyphoides, Heer

a (89) Temperate and subtemperate regions of the Northern Hemisphere, Andes, rare in West Indies and Madagascar

DISCUSSION OF THE TABLE OF DISTRIBUTION.

In attempting to compare and discuss a few of the more salient points which this table brings to light, it will perhaps be most convenient to consider the several groups of the systematic arrangement in their descending order from the primary subdivision into the two great series down to the ultimate subdivision into species. Preliminary to this a few of the leading facts need to be set down.

The whole number of species enumerated in the table is 1,540, of which 286 are Cryptogams and 1,254 are Phanerogams. The Cryptogams consist of 119 cellular and 167 vascular, and the Phanerogams of 115 Gymnosperms and 1,139 Angiosperms. The Angiosperms embrace 160 Monocotyledons and 979 Dicotyledons, and this last subclass is made up of 467 apetalous, 406 polypetalous, and 106 gamopetalous plants. These are the primary groups into which the vegetable kingdom is divided in the natural system, and, with the occasional exception of the last two, vegetable paleontologists almost unanimously adopt the order in which they have just been stated, which is also that of the table. They do this chiefly because it best represents the order in which these groups have appeared in the geological history of the earth, and their relative abundance in the several ascending strata. This, however, is true only as a general proposition, and may not hold in special cases, particularly when adjacent formations are compared. It cannot, therefore, be expected to prove literally true of the three formations we are here considering, nor to have any very great weight in determining the age of the Laramie group. Doubtless if we knew the entire flora of that group, and also the floras of the upper Cretaceous and the Eocene, such a comparison would have considerable weight and serve in large measure to fix the time at which the first of these floras flourished relative to that of the other two. But while we need not anticipate great results in this direction with things as they are, our table enables us to make this comparison, and it will be interesting, to say the least, to do so.

In comparing the leading floral elements of these three formations, however, it is evident that we cannot use the net figures as given above, on account of the occurrence of a considerable number of species in more than one of them, sometimes in all three. The number of such coincidences amounts in our table to twenty-four, making the gross entries in the three columns 1,564 instead of 1,540, and the former of these numbers must be taken as a basis of comparison. These slight additions will be scattered through the different groups, affecting them all more or less. The changes will not, however, at all vitiate the conclusions to be drawn. It is clear that the element to which we must attend is the proportion which the several vegetable groups bear to the total number from each formation, and that a comparison of these percentages in the same group for the three formations will afford us all the basis there is from which to draw conclusions.

The data may be condensed in the following form:

Systematic groups.	Laramie.		Senonian.		Eocene.	
	Number.	Per cent.	Number.	Per cent.	Number.	Per cent.
All plants	323	100.0	362	100.0	879	100.0
Cryptogams	48	14.9	97	26.8	143	16.2
Cellular	13	4.0	18	5.0	89	10.1
Vascular	35	10.9	79	21.8	54	6.1
Phanogams	275	85.1	265	73.2	736	83.8
Gymnosperms	18	5.6	43	12.2	58	6.6
Angiosperms	257	79.5	222	61.0	678	77.2
Monocotyledons	31	9.6	23	6.4	107	12.2
Dicotyledons	226	69.9	199	54.6	571	65.0
Apetalae	119	36.9	116	31.7	241	27.5
Polypetalae	84	26.0	66	18.2	263	29.9
Gamopetalae	23	7.0	17	4.7	67	7.6

An examination of these percentages shows that little light is thrown by them upon the relative age of the Laramie group. While in the Senonian, as theory would require, the Cryptogams have a higher proportion than in the other formations, it will be observed that they have a smaller proportion in the Laramie than in the Eocene, which is contrary to theory. This anomaly, however, is caused by the irregular representation of the cellular Cryptogams, which generally have increased with the later epochs and do not represent the waning types of the ancient floras. The vascular Cryptogams, however, do this, and it is to them that we must look for the confirmation of the theory, if it is to be confirmed. We find that it is here confirmed with sufficient accuracy, the Laramie occupying a position intermediate between the Senonian and the Eocene, though considerably nearer to the latter.

In the Gymnosperms we find the same anomaly as in the total Cryptogams, which in both cases is evidently due to the great predominance in the Laramie group of dicotyledonous forms. That group is, however, exceptionally rich in Monocotyledons, approaching the Eocene in this respect, while this type is meagerly developed in the Senonian. It is the great predominance of palms in the lower Laramie that has led Mr. Lesquereux to insist upon its Eocene facies, and this is certainly evidence not to be ignored. It is known that this type reaches its maximum development in the Eocene, and that to its predominance the special character of the Eocene flora is largely due. If, however, the Laramie group includes the Fort Union beds in one great deposit, with an extensive north and south range, its combined flora will certainly greatly reduce the percentage of these Eocene types, for we must recollect, and I hope soon to demonstrate this fully, that, so far as now published, the flora of the southern districts is given a wholly undue prominence and that of the northern remains as yet for the most part undescribed. Still, this is an anticipation which is out of place here, since the object of our present research is to inquire into the characteristics of the Laramie flora as hitherto published and made known.

The great profusion with which the Dicotyledons are represented in all these floras—amounting to considerably over half the species even in the Senonian, over two-thirds in the Laramie, and nearly two-thirds in the Eocene—makes this group of plants a somewhat more reliable term of comparison than any of the less abundantly represented types thus far considered. Whatever may be thought of the proper place of the Gamopetalæ, so sparingly preserved for us in the fossil state, it is universally admitted that the Apetalæ, or Monochlamydeæ, with their numerous amentaceous genera, furnished the earliest representatives of dicotyledonous vegetation, and that the forms with two floral envelopes (Dichlamydeæ) came later and form a higher type of plants. If we examine the percentages here, we find that the law holds true for the Polypetalæ and Gamopetalæ, which are the rising forms, or at least were so during all three of the epochs under consideration. The percentage is least in the Senonian, intermediate in the Laramie, and highest in the Eocene. In the Apetalæ, however, the maximum development appears in the Laramie instead of in the Eocene, which is not easily explained and probably will not continue to hold true with the more complete elaboration of that flora. These comparisons are with the total floras of the several groups, but perhaps a more interesting result will be obtained if we consider the Dicotyledons by themselves, and then find the relative proportions which the subdivisions bear to the whole in the three formations. Such a comparison will show that in the Laramie group the Apetalæ are 53, the Polypetalæ 37, and the Gamopetalæ 10 per cent. of the Dicotyledons; that in the Senonian the Apetalæ are 58.5, the Polypetalæ 33, and the Gamopetalæ 8.5 per cent. of the Dicotyledons; and that in the Eocene the Apetalæ are 42, the Polypetalæ 46, and the Gamopetalæ 12 per cent. of the Dicotyledons. On the theory that these types progressed in the order named and that the Laramie is intermediate between the other two formations, the relative number of apetalous species should diminish as we pass from the Senonian to the Eocene, which is the case, viz: Senonian, 58.5; Laramie, 53; Eocene, 42. The relative number of Polypetalæ, on the other hand, should rise with the age of the strata, and this we also find to be the case : Senonian, 33; Laramie, 37; Eocene, 46. The Gamopetalæ should also rise with the strata, but more rapidly. The figures are: Senonian, 8.5; Laramie, 10; Eocene, 12. These coincidences of fact with theory are interesting, and in view of the circumstance that they continue to hold from the Cenomanian below to the Miocene above,[1] they can scarcely be regarded as wholly without significance.

The advantage of comparing such large classes consists in the tend-

---

[1] See Fifth Annual Report of the United States Geological Survey, 1883-'84, pp. 449, 450. For similar data for the comparison of the floras of other formations, see table on pages 440 and 441.

ency of this method to eliminate the disturbing element of geographical distribution, which, as we shall soon see, is the chief obstacle to exact results in the consideration of genera and species. The species may all differ, the genera may be more or less local, even the orders may prevail in certain continents or hemispheres, but the relative predominance of such great types as the vascular Cryptogams, the Gymnosperms, or the Dicotyledons may depend chiefly upon the period in the history of their development, and, therefore, afford a measure of time which is as much more reliable as it is more rude and general than that afforded by the narrower groups of vegetation. Viewed in this light, the data thus far considered, while suggesting nothing more definite, may be fairly claimed to prove that the Laramie age was considerably later than that of the Senonian, and somewhat earlier than that of the Eocene flora.

In the classification of plants according to the natural method the next subdivision after the ones we have just considered is that into natural families or *Orders*. In certain large systematic works, it is true, an intermediate group is often introduced, usually called the *Cohort*, but it will not be necessary or convenient in the present case to treat this subdivision separate from the Order. In the cellular Cryptogams the classification is very unsettled, and the several groups receive different systematic values. The Fungi, Lichens, and Algæ are not always regarded as orders, but they are so rare in a fossil state and of so small importance from the chronological point of view that they may be conveniently so regarded here. Four species of Fungi, consisting chiefly of spots on dicotyledonous leaves (*Sphæria*, Hall) have been described from Laramie strata, while only one such has been reported from the Senonian and only two from the Eocene. The only lichen referred to any of these formations is an Opegrapha from the Laramie. Nearly a hundred species of supposed Algæ have been published from the three horizons, three-fourths of which are Eocene, embracing a large number of doubtful forms described (and often not figured) by Massalongo from Monte Bolca, etc. The Laramie furnishes only eight and the Senonian seventeen. Their diagnostic value may be set down as *nil*. Twelve Characeæ (all belonging to Chara) and four Muscineæ all from the Eocene, complete the cellular Cryptogams, which, for our present purpose, might as well have been omitted from the table.

The Filices, or Fern family, constitute an important order from the point of view of this discussion, furnishing 154 species. As the waning descendants of Carboniferous types that predominated throughout the earlier history of the globe, we naturally expect them to continue to bear in point of abundance some relation to the age in which they are found, the earlier to have precedence over the later. The assumed position of the Laramie group between the other two is borne out by this order, for, although a larger actual number of species occurs in the Eocene than in the Laramie, this number is less in proportion to the total of the two floras. The Laramie flora is 21 per cent. of the three combined floras, the Senonian is 23 per cent., and the Eocene 56 per cent. The quota of each, therefore, were they all of the same age, would be: Laramie, 32; Senonian, 36; Eocene, 87. It will be seen that the Senonian far exceeds its proportion, even assuming for it a considerably lower position. We are thus forced to see in the Senonian flora a much stronger Mesozoic facies than in either of the other groups. No family of plants brings out this fact more clearly than that of the Ferns, but it also speaks with equal authority upon the position of the Laramie below the recognized Eocene plant beds as thus far known.

The Rhizocarpeæ, Equisetaceæ, and Lycopodiaceæ can best be mentioned when we come to consider the genera, and we will now pass to the two gymnospermous orders, the Cycadaceæ and the Coniferæ. The Cycadaceæ, although they have barely survived into modern time, are, as is well known, a characteristic Mesozoic type of vegetation, having attained their maximum development in the Jurassic. They form an insignificant part of the Cretaceous flora and nearly disappear with the Tertiary. The only Laramie species rests upon a single specimen found at Golden, Colorado, and referred by Mr. Lesquereux to the genus Zamiostrobus. Yet seven species belonging to almost as many genera are recorded from the Senonian, again reminding us of the Mesozoic age of this flora.

We are thus brought to the consideration of the Coniferæ, which is one of the most important orders in the vegetable kingdom for the paleontologist. In the three formations under consideration this order has thus far yielded 107 species, of which 17 are found in the Laramie, 36 in the Senonian, and 58 in the Eocene, there being four coincidences. The even quota of each would be: Laramie, 23; Senonian, 26; and Eocene, 62. As the Coniferæ probably attained their maximum development in the middle Cretaceous, that is, earlier than any of the three epochs we are considering, the older of these epochs should show an excess over this quota and the younger a deficit. The Senonian shows such an excess and the Eocene such a deficit, but the Laramie also falls below even farther than the Eocene, which, in so far as the evidence of this order goes, gives it a more modern aspect than the Eocene.

Passing to the monocotyledonous orders, we find them, with the exception of the Palm Family, too small to afford any reliable criterion for the settlement of questions of age. The Naiadaceæ and Gramineæ are the only other orders at all approaching the palms, and both these display decidedly modern characteristics, compared with any of the types hitherto considered. If the palms reached their highest state and greatest abundance in Eocene time, the grasses did not probably attain this position before the close of the Miocene, and it may be doubted whether they have attained it at the present time. The same may be said for the Cyperaceæ and perhaps for the Naiadaceæ. The Liliaceæ and Sci-

tamineæ may have declined somewhat, as have more probably the
Aroideæ. It is at least evident that in considering the monocotyledonous orders we are confronted by a set of conditions the reverse of those
we met with in the ferns and the Coniferæ, viz: all our formations are
now below the period of maximum development of the group under
consideration, and the opposite results must be expected. These, in
fact, we find. The palms furnish 60 species, which, evenly distributed,
would give the Senonian 14, the Laramie 13, and the Eocene 33; but
the Senonian gets only 4, while the Laramie gets 17, the Eocene affording the remaining 39. In this important order, therefore, the Laramie
is about as fully represented as the Eocene, a fact which has been used
to its full extent in arguing for the Eocene age of the Laramie group.
If, however, we take the other monocotyledonous orders together, we
find that the Laramie (14) falls considerably more below its quota (21)
than the Senonian (19) falls below its quota (23), which might equally
be taken to argue its Cretaceous age.

In discussing the numerous dicotyledonous orders, we can only select
those which are most important, either from their abundance in the fossil state or from certain peculiarities or anomalies which they present.
As all trace of the earliest beginnings of this great subclass is still with-
held from human observation, it is difficult to describe the rise and de-
cline of its several subdivisions, but it seems probable that the mono-
chlamydeous forms were not only the earliest to appear, but that at the
period when we first make their acquaintance (the middle Cretaceous)
they had nearly attained their acme of growth and diversity. We then
find the large families Salicineæ, Cupuliferæ, Urticaceæ, and Laurineæ
in great profusion and highly developed, while many forms which are
now dichlamydeous, though they might not then have been so, had
already come upon the scene. In examining some of these large orders,
the principal question we have to ask is, Does their occurrence in
the Laramie group more nearly resemble that in the Eocene or in the
Senonian, or rather, assuming that the divergence of the Senonian and
Eocene; as known quantities, indicates difference of age, does the divergence of the Laramie from the Eocene indicate for that group an age at
all earlier than the latter? The comparison, as in former cases, must
be with even quotas and not with the actual figures. The Salicineæ
furnish 56 species to the three formations. The quota of the Eocene
would be 31, and we find 16; that of the Senonian should be 13, and we
find 14. An intermediate position would make the Laramie fall somewhat short of its quota (12). As a matter of fact it more than doubles
it (26). So far as this order would indicate, therefore, the Laramie
would be decidedly sub-Senonian. This is due to the great predominance of the genus Populus in the Laramie group, of which more will
be said hereafter.

The Cupuliferæ furnish 146 species. Of these the Eocene has 58,

a number about one-third below its quota (82), while the Senonian has
52, a number as much above. The Laramie occupies a strictly intermediate position, yielding 36 species, or five more than its quota. In the
Urticaceæ the Laramie deviates more from the Eocene than does the
Senonian and in the same direction as in the Salicineæ, while in the
Laurineæ the deviation is again intermediate. In the Juglandeæ we
again have the Laramie showing an exaggerated Mesozoic tendency.

We thus see that none of the apetalous orders give the Laramie the
same position, from this numerical point of view, as the Eocene, all
placing it lower and either intermediate between the Eocene and the
Senonian or below the latter.

The principal polypetalous orders are the Araliaceæ, the Myrtaceæ,
the Rosaceæ, the Anacardiaceæ, the Sapindaceæ, the Rhamneæ, the
Celastrineæ, the Sterculiaceæ, and the Magnoliaceæ. They are much
more decidedly Eocene in aspect than the apetalous orders, but less so
than they appear with the proportionally large figures in that column.
In fact, the Eocene generally only slightly exceeds its quota for the three
groups after equalization as explained above, and in the Rhamneæ and
Magnoliaceæ it falls below it. A careful inspection of these nine orders
shows that in two cases (the Rosaceæ and the Sterculiaceæ) the Laramie
holds an intermediate place between the Eocene and the Senonian, that
in four cases it holds a place below the Senonian, while in three cases
(the Anacardiaceæ, Sapindaceæ, and Magnoliaceæ) its position is indicated as slightly higher than the Eocene.

The gamopetalous orders are small and their indications are readily
deduced from a casual inspection of the table. The two largest, the
Ebenaceæ and Caprifoliaceæ, consist entirely of the two genera, Diospyros and Viburnum, respectively, and can be treated under the head of
genera. Taking all the gamopetalous orders together, the Laramie is
seen to occupy an intermediate position between the Senonian and the
Eocene.

In examining the orders represented in the three formations under
consideration, especially the smaller orders, a marked tendency is visible toward the confinement of entire ones to one formation. This is
due to geographical peculiarities, a characteristic which, when we come
to study the genera, can be no longer ignored.

We are now prepared to consider our subject from the point of view
of the *genera*, and before going further it will be necessary to point
out some of the difficulties of this method. In vertebrate paleontology
the genera are nearly all extinct, and therefore the paleontologist
may here legitimately employ his genera as reliable data for the
determination of the age of the formations to which they are confined.
In vegetable paleontology this is by no means the case. Of the 354
genera represented in the three formations only 165 are extinct, and

many of these are so similar to living genera as to be designated by
the same names with modified terminations, such as *ites, opsis, etc.*, and
such forms are, with better material and more careful study, being constantly made to take their places as true living genera. The vertebrate
paleontologist, therefore, deals with genera as the paleobotanist does
with species, and in fact, as is well known, in this department of zoology
the term "genus" is given a much more limited meaning than it is in
botany, and a rank not far above that of "species" among plants. This
is doubtless in great part necessary, and due to nature having drawn
classificatory lines, so to speak, at somewhat different points in different
scales of being. But it is clear that the paleobotanist cannot compare
his genera as the vertebrate paleontologist compares his for the settlement of questions of geologic age. It is, however, true that certain
genera which flourish at the present day predominate in certain formations and are rare or absent in others of later age, so as in a true sense
to be characteristic of such formations. This does not prove that they
subsequently dwindled away and then revived at a still later date,
although this might, and probably sometimes does, occur. But the explanation is that several beds of different age are usually in different
parts of the world, and the flora of the globe in past time, as at present,
has sustained different types of vegetation at different points on its
surface. Or, if the beds are nearly over each other, *i. e.*, not far separated geographically, the predominance of certain genera in lower that
are rare or absent in higher strata must be explained on the hypothesis
of migration or by supposing that the nature of the country at the two
points was very different at the time of the respective deposits. It
thus comes about that when we speak of the Laramie flora we refer to a
definite geographical area at a definite period of time, and when we speak
of the Eocene flora we mean the beds occurring at the localities named
on our table and a few others grouped together in the last column but
one. If the reader will take the trouble to inspect the columns of the
table in which the Senonian species are set down he will find that a very
marked distinction exists between those of Europe on the one hand and
those of America and the Arctic regions on the other, and that the latter resemble much more closely those of the Laramie group. This is
entirely because they are in nearer geographical relationship with them.

But it must not be forgotten that genera are capable of great modifications without rendering a change of name necessary, and the prac-
tice among paleobotanists has been to crowd everything into living genera that they will contain without doing violence to their accepted at-
tributes. Therefore, an Eocene or a Cretaceous genus, though still living,
may embrace forms widely divergent from those now recognized under
the same name, so that such genera may really be characteristic of those
formations as strictly as though they had become extinct at their close.
The principal interest, therefore, centers upon these characteristic

genera, by which term we do not here mean either that they are extinct genera, or that they do not occur in higher strata (*e. g.*, Miocene), or in lower (*e. g.*, Cenomanian), or that they are wholly excluded from either of the three formations, but simply that they predominate in some one relatively to the other two.

As already stated, the whole number of genera represented in the three formations is 354. Of these, 32 are confined exclusively (so far as these formations are concerned) to the Laramie group, 62 to the Senonian, and 155 to the Eocene; 49 are common to all three formations, 6 are found in the Laramie and Senonian, but not in the Eocene, 23 are found in the Laramie and Eocene and not in the Senonian, and 27 are absent from the Laramie and found in both the other formations. The number found at only one horizon is therefore 249, the number occurring at two horizons 56, and the number at all three 49. The number ranging from the Senonian to the Eocene, and therefore, regardless of the Laramie, certainly belonging to both Mesozoic and Cenozoic time, is 76.

The discussion of the genera may be conveniently separated into two parts, one of which shall be devoted to the consideration of the evidence in favor of synchronism, and the other to the subject of geographical distribution. The first of these subdivisions will have nothing to do with any of those genera which are, in the sense here employed, characteristic of any one of the three formations, but must be confined to those that are common to two or all three. Such genera, moreover, as are nearly equally represented in each of the three formations can have no weight in establishing the affinity of the Laramie with the one rather than the other, and must also be excluded from our primary comparisons. A further exclusion must be made of those genera which are common to the Senonian and the Eocene but absent from the Laramie, since both these formations are treated as known quantities, and comparison of their common elements could lead to no new results. We are therefore really reduced to such genera as are either confined to the Laramie and Senonian or to the Laramie and Eocene, or are so nearly thus confined as to be fairly characteristic of the two. In deciding such cases we may also properly exclude very small genera, such for instance as are represented by only one or two species in each formation, unless these species be specially diagnostic or very abundant; but we must not at any time lose sight of the fact that it requires about two and a half species in the Eocene to have the same weight as one in either of the other formations.

After carefully scanning the table, I have selected such genera as I think fairly illustrate this point, and they may be set down in their systematic order in two opposing columns, with the number of species belonging to each:

Laramie and Senonian.				Laramie and Eocene.			
Genera.	L.	S.	E.	Genera.	L.	S.	E.
Zamiostrobus	1	1	....	Halymenites	3	....	1
Abietites	2	1	....	Caulinites	2	....	6
Taxites	2	1	....	Sabal	4	1	9
Sequoia	6	9	4	Flabellaria	2	1	4
Taxodium	3	3	1	Alnus	2	1	6
Phragmites	3	1	....	Rhus	5	1	9
Populus	23	9	7	Sapindus	4	1	5
Juglans	8	2	2	Vitis (?)	5	....	3
Platanus	8	2	....	Zizyphus	5	1	...
Cornus	4	2	2	Celastrinites	2	....	4
Acer	3	2	2	Grewiopsis	2	....	6
Rhamnus	12	2	3	Dombeyopsis	4	....	14
Paliurus	2	1	1	Magnolia	6	2	9
Fraxinus	2	1	1				
Viburnum	15	4	2				

We thus have fifteen genera belonging to the first class and thirteen to the second. Both lists would admit of reduction, but some good reason can be urged in each case for retaining it.

We may examine these several characteristic genera somewhat in detail. Beginning with the first list we find a single species of Zamiostrobus in the Laramie and in the Senonian. The latter occurs in the Gosau formation at St. Wolfgang, Austria, which is now believed to be definitely settled as upper Cretaceous. The Laramie plant is of a somewhat doubtful character, but is clearly cycadaceous. It was found at Golden, Colorado, lying on the surface in the vicinity of Laramie beds, and is believed to belong to that formation. The genus, like all fossil cycadaceous genera, is strongly Mesozoic, being found as low as the Oolite.

Abietites, two species of which occur in the Laramie, one being found in both the lower and the upper district, is one of the most ancient of the typical coniferous forms, being found all the way from the Wealden to the Miocene, except in the Eocene, where it is thus far absent. The only Senonian species comes from the Harz district.

The form distinguished as Taxites seems to belong to the northern portion of the western hemisphere, the two Laramie species being reported from British America, and the Senonian species from the beds of Patoot, Greenland. A true Taxus occurs in the London clay, and this seems to be a geographical variety.

No coniferous form is more abundant in the Laramie than Sequoia, six species of which are distinguished. Of the nine species from upper Cretaceous strata all but one are found in the western hemisphere. This furnishes an excellent illustration of the extent to which certain types persist with modification in the same or adjacent territorial areas. There is no doubt that should upper Cretaceous beds be found within the United States these forms will occur as the direct ancestors of the Laramie species. Their rarity in the Old World is seen also to be a fact of geographical and not of geological significance, for it is true of both the Cretaceous and the Eocene.

The genus Taxodium, two of the species of which are so abundant in the Laramie, Senonian, and Miocene, is curiously scarce in the Eocene, and therefore claims a place in our first column.

It is in the Gymnosperms, therefore, that those characters appear which give to the Laramie flora such a strong Cretaceous facies. We find this quite otherwise in the next group, the Monocotyledons. Only in one genus (Phragmites) of this subclass do we find the Eocene wanting. This genus occurs abundantly in the Laramie, and the only Senonian species reported is from the Pacific coast of America, so that it seems that in pre-Miocene time the type was confined to the western hemisphere.

It is, however, among the Dicotyledons, and chiefly in the Amentaceæ, that the most notable examples occur to show the similarity of the Laramie to the Senonian flora, and also its unique character as compared with any other formation. Its 23 species of Populus form one of the greatest of its anomalies, and stamp it with a special character. The nine species of the Senonian cause that formation to partake somewhat of this character, but when we see that all but two of these come from the Vancouver beds or from Greenland we see that this is a distinctly American type.

The genus Juglans, with its eight Laramie, one Vancouver, and one Patoot species, is of special interest in the light of the numerous forms of Carya and Juglans which persist in the American flora. The fossil forms of Juglans may well have been the ancestors of our hickories as well as of our walnuts.

Neither of the two last-named genera, however, can claim as great a share of our interest as does the genus Platanus. With its eight Laramie and two Greenland species, and its entire absence from the Eocene, it seemed to constitute in pre-Miocene time one of the characteristic vegetable types of America.

Passing over the two polypetalous genera, Cornus and Acer, which in like manner belonged during this epoch almost entirely to the west, we come to Rhamnus, with twelve Laramie species; one of the Senonian species is also western (Patoot). Paliurus is an allied genus and is similar in its range to Rhamnus.

Of gamopetalous genera, Fraxinus, though small, belongs to the class we are considering, while Viburnum is, next to Populus and Platanus, the largest and most characteristic of that class. With fifteen species in the Laramie, four in the Senonian, and the two Eocene species from the lowest beds of that age, it seems to be a very ancient type, and one which goes far to separate the Laramie flora from that of the Eocene.

If there were no cases which could be cited to offset this array of evidence, it might seem that no two floras could be more distinct than those of the Eocene and the Laramie, but as we pass rapidly down the opposite column we shall see that there certainly are some bonds of union.

It was long maintained that the peculiar fucoids called Halymenites were characteristic of the Eocene, being so abundant in the Flysch of Switzerland, and their presence in the Laramie strata was put forward as a proof of the Eocene age of that group, but they are now known to occur in the Cretaceous, though absent from the Senonian beds, and as low as the Jurassic. They also extend upward to the Miocene.

The two species of Caulinites from the Laramie differ widely from those of the Paris Basin, but probably belong to that type of plant and in so far assimilate the Laramie to the Eocene flora. It is, however, the palms that have been chiefly relied upon to establish the Eocene character of the Laramie. The evidence here must be admitted to be strong, and their absence from the Senonian beds serves to add to its force. The Eocene was the age of palms. The numerous fruits referable to that family found in the London clay and also at Monte Bolca, constitute one of the leading features of the flora of that epoch, and these are in a manner paralleled in some parts of the Laramie, notably in the tufa beds at Golden, by the many nut-like bodies which Mr. Lesquereux has designated by the term *Palmocarpon*. But aside from these, and probably from the same trees that bore them, we have four species of Sabal and two of Flabellaria represented by leaves in the Laramie flora, though nearly all these palms are found in the lower districts. It is only this lower Laramie that has been claimed as Eocene, and if we restrict the term to this flora its affinity to that of the European Eocene is greatly strengthened.

The genus Alnus is well represented in the Eocene, especially in the Paleocene, and one abundant species is found in the Laramie group. The Senonian species is from Greenland and may have been the progenitor of the wide-spread arctic form *A. Kefersteinii*, Göpp., so celebrated in the Miocene beds of the North.

The Marquis Saporta finds eight species of Rhus in the gypsum beds of Aix in Provence, and the genus also occurs in all the Laramie horizons. The type therefore is common to the two formations and serves to assimilate the two floras. The one Senonian species is from the Quedlinburg beds.

Sapindus predominates in the Fort Union group and in various Eocene localities, and in so far tends to identify the upper Laramie with the Eocene; but such evidence is very feeble. Vitis is a strong Laramie genus, but it occurs sparingly in the Eocene. It therefore scarcely belongs in this list. Zizyphus differs from the other two prominent rhamnaceous genera, Rhamnus and Paliurus, in extending into the Eocene. It is a fair representative of the class we are now considering that indicate a resemblance between the Laramie and the Eocene floras.

The Celastraceæ are highly characteristic of the Eocene, and one form which has been distinguished as Celastrinites is found in the Laramie. The Eocene species of this genus are all from Sézanne, and furnish another evidence of the truth of Mr. Lesquereux's statement in his "Cretaceous and Tertiary Flora" that the flora of the Laramie resembles that of Sézanne more closely than it does that of the Eocene proper. A still more striking illustration of the same fact is found in Grewiopsis, which is the Paleocene form of the Miocene genus Grewia, also occurring in the Laramie.

Dombeyopsis is one of the best marked Eocene genera, but it is almost exclusively confined to Monte Bolca. Its occurrence in the Laramie group is a singular fact and one that has often been brought forward in support of the Eocene age of that group.

The Magnoliaceæ are a very ancient type of plants, species of Liriodendron being abundant in the Cenomanian. The genus Magnolia, which occurs in the upper Cretaceous beds of the Peace and Pine Rivers in British America, is abundant in both the Laramie and the Eocene. It is simply a persistent type.

We have thus rapidly run over the evidence furnished by these two classes of genera for and against the view that the Laramie flora bears such a resemblance to the Eocene flora as to suggest the substantial synchronism of the two series of deposits. It is perhaps best to leave the reader to form his own judgment as to the result, but in the light of former discussion of this question the caution against mistaking horizontal for vertical distribution, may not be out of place. In the great majority of cases, as has been pointed out under each genus, the types persist through different ages in the same or adjacent parts of the the world, and the absence of Laramie types in the Eocene, and *vice versa*, is due to the wide geographical separation of the beds of the two formations. Closer study of the table will show that most of the European genera can be traced from the Cenomanian up to the Miocene of that continent, while most of the American genera can be traced from the Dakota group up to the Miocene of Alaska and Greenland. That some genera should be common to both hemispheres was to be expected, but that these distinctly argue either the Eocene or the Cretaceous age of the Laramie beds cannot be reasonably maintained.

This is the proper place, before descending to specific details, to consider this interesting subject of geographical distribution in its relation to the present plant life of the globe. The present distribution of vege-

table forms upon the earth's surface, as all know, is very varied, and
several learned and largely successful attempts have been made to
trace the lines of migration of plants during their long and often tor-
tuous pilgrimages since Miocene times, driven as they have been by
successive alterations of climate, of sea and land surface, and of mountain and plain. But we have seen that the flora of the globe, even as
early as the Cretaceous, was far from uniform at all points, and that
that of the eastern and western hemispheres in late Cretaceous and
early Tertiary time was widely different. We now find that the degree of change since those epochs has been different at different points
and far greater in Europe than in America. The data contained in the
foot-notes to our table enable us to demonstrate this, and also to show what
parts of the globe contain at the present time the leading elements of
each of the fossil floras under consideration. If we exclude those genera which are abundant in all three formations, and take only those that
are either wholly or principally confined to one of them, we shall perceive that the greater part of the properly Laramie genera are represented to their fullest extent in the present flora of North America or
eastern Asia, though many belong to the warmer parts of America,
and to India. On the other hand we are struck by the very large number of Australian and African forms in the Eocene flora. The Pro-
teaceae and Myrtaceae abound in the Eocene as do the Leguminosae,
the latter chiefly of South African types. We also find that the Senonian flora must be separated into two classes, those from British America and Greenland falling into the same general geographical group as
those of the Laramie, while those of the European beds resemble the
Eocene flora in this respect. I had intended to elaborate these choro-
logical features more at length and to give a detailed analysis of the
three floras from this point of view, but space will not admit of this in
the present paper, and as all the data for such an analysis exist in the
preceding table of distribution the work of compilation may be left to
such as are particularly interested in this feature of the discussion. The
results upon their face fully bear out the statement already made that
the flora of the Laramie group furnishes evidence of having descended
more or less directly from that of the Cretaceous of this continent, and
in many cases the lines of descent can be traced through the upper, or
Senonian beds to those of the Dakota group, or American Cenomanian.

We are now prepared to compare the three floras under consideration from the usual point of view of their specific relationships, and if
the treatment of this part of the subject is brief it is for the very reason that it has already been largely accomplished by others. Still, as
already remarked, Mr. Lesquereux only embraces the flora of the lower
districts, exclusive of Carbon and Evanston and a few Upper Yellowstone localities, in his Laramie group, while our table combines all
these beds with the entire Fort Union deposit of the Upper Missouri

and Lower Yellowstone. As these latter were, and by many are still, regarded as Miocene, and certainly contain a flora differing in many respects from the rest, the general complexion of the whole will be considerably modified by including them.

By inspecting the table we observe that only a single species, *Sequoia Langsdorfii*, is common to all three of the formations. This species is generally northern in the western hemisphere, but it is found in the Laramie at Black Buttes, in the Fort Union group, and in the northern extension of this latter in British America. It also occurs in the Cretaceous deposits of Nanaimo, Vancouver Island, and in the Senonian beds of Patoot, Greenland. Professor Gardner finds it in the Eocene deposits of the Isle of Mull, and Massalongo enumerates it in his Miocene flora of Senegal.

Only one other Laramie species, *Ginkgo polymorpha*, is found in any of the Senonian beds, and this occurs also at Nanaimo. Its Laramie locality is the place near Fort Ellis in Montana designated as "six miles above Spring Cañon," which we have seen reason to regard as a western member of the great Fort Union deposit.

The number of Laramie species that also occur in the Eocene as defined in the table is quite large, amounting in all to thirteen or fourteen. Seven of these are confined to these two formations, which might afford strong prima facie evidence of the close affinities of the Laramie and Eocene floras. This evidence, however, is greatly weakened when we perceive that of these seven four occur in the supposed Eocene beds of Mississippi and not in any of the Old World deposits. This is certainly strong proof of the close relationship of these Mississippi beds to those of the Laramie, as well as of their similarity of age, but it is more interesting as showing that in those early times one great homogeneous flora stretched all the way across the North American continent, and that similar forests fringed the waters of the Gulf of Mexico during their southward retreat, and those of the Laramie Sea as it shrunk to the proportions of inland lakes. The difference of time between the two deposits, though it might have been great, was not sufficient to alter the specific identity of these four forms and doubtless of very many others, while in other cases the Laramie species may represent the ancestors of the Eocene species found or to be found in the more eastern deposits. These species are, *Sabal Grayanus*, *Populus monodon*, *Magnolia Hilgardiana*, and *M. Lesleyana*, all of Lesquereux. All except *Magnolia Hilgardiana* occur only in the typical Laramie deposits of the more southern districts, but this species has now been reported also from the Yellowstone Valley, which, of course, relegates it to the Fort Union group.

The other three Laramie species which are otherwise confined to the Eocene are *Halymenites minor*, found in the Flysch of Switzerland, *Ficus Dalmatica*, found in the supposed upper Eocene beds of Monte Promina in Dalmatia, which some authors place higher, and *Sterculia modesta* of

6 GEOL——34

Saporta (not of Heer) found at Sézanne. These three Eocene localities represent the highest and lowest Eocene, and fairly exhibit the degree of homotaxy subsisting between these formations.

The remaining six species that occur in the Laramie and the Eocene, possess less force in this direction from the fact that they are all found in other and higher formations also. Most of them are plants that are abundantly represented in nearly all the more recent deposits, such as *Taxodium Europæum*, found all the way from the Middle Bagshot of Bournemouth to the Pliocene of Meximieux, *Ficus tiliæfolia*, *Laurus primigenia*, and *Cinnamomum lanceolatum*, abundant in nearly all the Oligocene and Miocene beds of Europe. *Quercus chlorophylla* occurs in the Mississippi Tertiary as well as at Skopau in Sachs-Thüringen, and is also abundant in the Miocene, and *Ficus tiliæfolia* is found in the Green River formation at Florissant, Colorado. The only other species belonging to this class is *Goniopteris polypodioides*, which occurs at Monte Promina and in the Miocene of Rivaz. *Alnus Kefersteinii*, once reported from Aix in Provence, is considered doubtful, and should probably be excluded from the list of Eocene plants, but it is found in the American Eocene of both Florissant and Green River. In the Laramie it is only known from the Evanston coal beds, and is most abundant in the arctic Miocene of Alaska, Spitzbergen, etc., but it is also common in the Miocenes of Northern and Central Europe.

This is all that can be said in favor of the Eocene character of the Laramie flora, and were it not capable of being further weakened, the case might be regarded as somewhat stronger than that of the genera; but there still remain many important considerations which affect the legitimacy of some of these facts. For example, we have seen that fourteen species altogether occur in the Laramie and the Eocene; but the number occurring in the Laramie and formations higher than Eocene is sixty-two. Thirty-five of these are confined to the Laramie and Miocene. Two (*Diplazium Mülleri* and *Flabellaria Zinkeni*) are confined to the Laramie and Oligocene, while twelve occur in Laramie, Oligocene, and Miocene strata. These species are by no means confined to those that have only been found in the northern districts, but, as any one can see by examining the table, they come largely from the typical beds, and include such species as *Sabal Campbellii*, *Salix integra*, *Betula gracilis*, *Ficus asarifolia*, *Rhamnus alaternoides*, etc.

It would certainly be very unsafe from this to argue that the lower Laramie is Miocene. With such a vast flora as the Miocene, numbering as it does (including the Oligocene beds) nearly 4,000 species, it is reasonable to expect that as many Laramie forms as are found common to the two formations (about 1½ per cent.) should persist nearly unchanged from one epoch to the other. As a matter of fact, a much larger percentage of forms thus persists where the two deposits occupy nearly the same geographic area. Some four or five of the Laramie species are still found in the living flora, most of them in North America, un-

changed, so far as can be judged by the organs (chiefly appendicular)
that have been found in the fossil state. The two species of hazel, and
also the sensitive fern from the Fort Union deposits regarded by Dr.
Newberry as identical with the living forms, must be specifically so referred until fruits or other parts are found to show the contrary. The
bald-cypress of the Laramie swamps seems not to have been specifically distinct from that of the swamps of the Southern States, and, as I
shall soon show, forms of the Ginkgo tree occur not only in the Fort
Union beds, but in the lower Laramie beds at Point of Rocks, Wyoming
Territory, which differ inappreciably except in size of leaf from the living
species.

To the strong evidence against the Eocene age of the Laramie group
afforded by the persistence of so many of its types into periods much
more recent than Eocene may perhaps be added evidence equally ad-
verse but of the opposite nature. A few Laramie forms occur in Cretaceous strata. *Sequoia Langsdorfii* is found, as we have already seen,
in the Cretaceous of both British Columbia and Greenland, and *Ginkgo
polymorpha* in the former of these localities. *Cinnamomum Scheuchzeri*
occurs in the Dakota group of Western Kansas as well as at Fort Ellis.
Sir William Dawson detects in strata regarded as Laramie by Prof. G.
M. Dawson, of the Geological Survey of Canada, a form which he considers to be allied to *Quercus antiqua*, Newby., from Rio Dolores, Utah,
in strata positively declared to be the equivalent of the Dakota group.

Besides these cases there are several in which the same species occurs in the Eocene and the Cretaceous, though wanting in the Laramie. *Cinnamomum Sezannense*, of the Paleocene of Sézanne and Gelinden, was found by Heer, not only in the upper Cretaceous of Patoot, but
in the Cenomanian of Atane, in Greenland. *Myrtophyllum cryptoneuron*
is common to the Paleocene of Gelinden and the Senonian of Westphalia, and the same is true of *Dewalquea Gelindensis*. *Sterculia variabilis* is another case of a Sézanne species occurring in the upper Cretaceous of Greenland, and Heer rediscovers in this same Senonian bed the
Eocene plant, *Sapotacites reticulatus*, which he originally described from
Skopau in the Sachs-Thüringen lignite beds.

Before commencing this discussion from the point of view of specific
relationship it was remarked that it would differ from that just closed,
where the subject was treated from the point of view of generic relationship, in dealing with geological, or time relations, rather than with
geographical, or space relations. But we have already seen that the
latter considerations could not be kept wholly out of view, and we shall
now see that they really form a very important part of this mode of treatment, if it is to be made at all complete. Of the seven species confined
to the Laramie and Eocene it was seen that four were also confined to
this continent. This anomaly arose from having placed the Mississippi
Tertiary in the last column of Eocene localities. But the Green River
group, which is by most geologists regarded as the Eocene of Western

America, was purposely left out of the body of the table, for reasons
which have been stated. A column, however, was employed to record
the occurrence in that group of species belonging to any of the three
formations. An inspection of this column shows that 24 species are
common to the Laramie and the Green River groups. Admitting this
to be Eocene, as well as the Mississippi Tertiaries, we have 26 species
common to the Laramie and American Eocene against 10 that are
common to the Laramie and European Eocene; this notwithstanding
that the American Eocene embraces less than a third as many species
as the European.

We may carry this analysis further. There are 39 species common
and confined to the Laramie and the Miocene (inclusive of the Oligocene). Of this number 24 are found in the American Miocene. Three
others occur in the arctic flora of Spitzbergen, Siberia, and other localities not in the western hemisphere, but the complete unity of the arctic
Miocene, and its almost total dissimilarity from the Miocene of Europe,
fairly warrant their addition to the American flora. Fifteen of these
are not found at all in the Miocene flora of Europe. This is surprising
when we consider how very small this combined North American and
arctic Miocene flora is compared with that of Europe.

If we now divide the Laramie species that are also found in other
formations and localities into two classes, one of which shall embrace
all those occurring in American beds other than Laramie and the other
those occurring in no other American strata than those of the Laramie, we shall have 55 such species out of a total of 80, 30 of which are
confined exclusively to the western hemisphere. The significance of
these figures, let me repeat, is greatly increased when we consider in
the same connection the magnitude of the European Tertiary flora, as
compared with that of America.

We are thus brought once more face to face with the fact that while the
floras of Europe and America differed widely in character during late
Cretaceous and Tertiary time, the beds of different age in each, respectively, contained floras resembling each other to such an extent as to
warrant the conclusion that the later ones had descended from the
earlier without more than the natural amount of modification. When,
therefore, we couple these facts with those presented above as to the
relationships of the fossil to the living flora of the globe (where it appeared that the American fossil flora resembles that of eastern North
America and southeastern Asia, while the European fossil flora resembles that now found in Australia and the eastern half of the southern hemisphere generally), we must conclude that some great disturbing agencies have been at work since Miocene times which have caused
extensive migrations and profound alterations in the plant life of the
globe. It is no part of my purpose at present to discuss this problem,
and I need scarcely say that it is to the influence of a series of great
fluctuations of temperature, causing glacial epochs, that these changes

are principally attributed, and that a thorough study of the living flora in comparison with the Tertiary flora not only bears out this conclusion to a remarkable degree, but renders it possible to trace many of the lines of migration and to fix with some precision both the space and the time relations of glacial phenomena.

We may now briefly revert once more, and for the last time, to the question of the age of the Laramie group, in so far as this is indicated by the similarity of its flora to that of other formations. Thus far I have confined myself to the published flora of that group in order to ascertain how the case stood at the close of the prolonged discussion which has been outlined relative to its age, in which discussion Mr. Lesquereux has had the last word in his recent great work on the Cretaceous and Tertiary Floras of the West. But I should admit that I was led to consider this side of the subject by the occurrence in my own collections from both the northern and the southern districts—in the Lower Yellowstone Valley and along the Upper Missouri, at Golden and other points in Colorado, at Carbon, Black Buttes, and Point of Rocks, Wyoming, and at other localities—of new forms, some of them unique and remarkable, but some bearing a striking resemblance to, or identical with, forms already figured from other localities whose stratigraphical position is definitively settled. While some of this latter class have a Miocene aspect, as does the Fort Union flora in general, there are others embodying the characters that are usually associated with the Cretaceous flora. As already remarked, it is too early for me to discuss these forms fully or in detail, although some of the more remarkable or representative ones are figured in the illustrations at the close of the paper. At present I can merely call attention to some of these forms of Cretaceous aspect, as showing that the more familiar we become with this flora the more closely we find it linked with the Cretaceous floras below it, and particularly with those of America.

There seems some reason to believe that we now have in Fort Union strata a somewhat modified representative of the hitherto exclusively Cretaceous genus *Credneria*, so long known from the upper Cretaceous beds of Blaukenburg, in the Harz Mountains, since found in other European strata of the same or earlier age, and now added by Heer to the middle Cretaceous flora of Greenland. *Credneria* is the original form upon which have since been erected the additional genera of the group *Ettingshausenia*, *Protophyllum*, and *Aspidiophyllum*. These are all characteristic Cretaceous genera, Credneria and Protophyllum being found both in the Senonian and the Cenomanian, and Aspidiophyllum being confined to the Dakota group. Our form (Plates LVII and LVIII) differs somewhat from all that have thus far been described, and may be sufficiently divergent to warrant the establishment of a new genus, or it may be necessary to refer it to some other genus, but its resemblance to *Credneria* is sufficient at least to make it a strongly Cretaceous type, and should its relationship to that genus be finally settled it must certainly possess weight in the general problem of geologic age. It is also noteworthy that this form comes from the Fort Union beds on the Lower Yellowstone, and from one of the highest strata of this formation that are represented in that section.

There occur in the collections a large number of querciform leaves, probably for the most part referable to the Cretaceous genus *Dryophyllum*, established by Debey as the receptacle for the numerous archaic oaks which he found in the iron sands of Aix-la-Chapelle. Until quite lately this genus was very little known, and chiefly from specimens furnished by him to different museums in Europe, but within the past two years he has published a small pamphlet with one plate, illustrating several of the forms.[1] The material seemed rather obscure and fragmentary, and the figures are very rude, but they enable us to gain a better idea of the limits of the genus than was otherwise possible. We have from the Laramie group forms closely allied to several of Debey's species of Dryophyllum, such as *D. Eodrys*, *D. gracile*, *D. cretaceum*, *D. Aquisgranense*, etc., although it is hardly probable that any of these species actually flourished in America.

There can scarcely be a doubt that we have in Figs. 8 and 9, Plate XL, the Cretaceous species *Platanus Heerii* of the Dakota group and arctic Cenomanian strata. Compare, for example, fig. 1 of plate vii, in the sixth volume of Heer's "Flora fossilis arctica," Part 11, Cretaceous flora of Greenland.

Several forms of Hedera have a Cretaceous aspect, and it is quite probable that *H. primordialis*, Heer, from the Greenland beds at Atane, may be represented by our Fig. 4, Plate XLVIII.

In Fig. 1, Plate LX, we have a form which, for so much of the leaf as is present, resembles the figures of similar portions of Heer's *Populus Stygia* (Fl. foss. arct., Vol. III, Kreidefl. v. Grönland, plate xxix, fig. 10; Vol. VI, Abth. II, Kreidefl. v. Grönland, plate xvii, figs. 5, 7; plate xxxix, fig. 5). But for the great resemblance to these figures. 1 should have certainly regarded it as a Liriodendron, and notwithstanding this resemblance I am inclined to refer it to that genus. But Liriodendron is rather a Cretaceous genus, although the broad-leaved forms like this occur also in later strata and form the type to which the living species belongs.

I have not mentioned the singular cryptogamous form that was collected both at Iron Bluff and at Barns's Ranch, although I am now convinced that it is a Cretaceous form, because up to the time when it was necessary to submit this paper it had not been sufficiently studied and the drawings were incomplete; but upon careful comparison I am satisfied that it is the same plant that is figured by Dawson in his paper in the Transactions of the Royal Society of Canada (plate i, fig. 3) as

---

[1] Sur les feuilles querciformes des sables d'Aix-la-Chapelle, par le Dr. M. Debey, d'Aix-la-Chapelle. Extrait du Compte rendu du Congrès de botanique et d'horticulture de 1880. Deuxième partie. Bruxelles, 1881.

Carpolithes horridus. To the parts represented there our specimens add the complete rays showing what is probably the spore-bearing portion at their extremities.

Other Cretaceous forms might be mentioned, but the above-named types are sufficient to show that the flora of the Laramie group certainly possesses a strong Cretaceous facies, and in very many respects agrees with that of the Senonian or highest member of that formation wherever this is known to contain vegetable remains. I do not wish to be understood as arguing that the Laramie is a Cretaceous deposit, but rather against the view maintained by Mr. Lesquereux that it is necessarily Eocene. I am still free to admit that, so far at least as the Fort Union group is concerned, the flora is closely in accord with that of the European Miocene, in which nearly all its genera and many of its species are represented; and but for the occurrence of these anomalous, archaic forms, which become more and more frequent as the material for study increases, it would be impossible to deny that the flora at least was Miocene. In this, however, one fallacy should be avoided, which is, I think, the one that so strongly biased Professor Heer in favor of referring new and imperfectly known floras to the Miocene. The immense number of fossil plants that are known from that formation — over 3,000 species — greatly increases the chances of finding the analogue of any new form among its representatives. While, for example, there are probably many more Laramie forms that have nearer allies in the Miocene flora than in that of any other age, still, relatively to the number of Miocene species, the Eocene or Senonian types would outweigh them. But the same canon must be applied in comparing the Laramie with these latter. If the relationships were about equal we should require a larger absolute number of Eocene forms, because the Eocene flora is larger

Taking all these facts into consideration, therefore, I do not hesitate to say that the Laramie flora as closely resembles the Senonian flora as it does either the Eocene or the Miocene flora. But again, I would insist that this does not necessarily prove either the Cretaceous age of the Laramie group or its simultaneous deposit with any of the upper Cretaceous beds. The laws of variation and geographical distribution forbid us to make any such sweeping deductions. With regard to the first point it is wholly immaterial whether we call the Laramie Cretaceous or Tertiary, so long as we correctly understand its relations to the beds below and above it. We know that the strata immediately beneath are recognized upper Cretaceous and we equally know that the strata above are recognized lower Tertiary. Whether this great intermediate deposit be known as Cretaceous or Tertiary is therefore merely a question of a name, and its decision one way or another cannot advance our knowledge in the least.

With regard to the synchronism, as already remarked, it would certainly be interesting and important if we could know with certainty what other deposits on the earth's surface were being made at the same time with those of the Laramie. But we have seen that this cannot be known for any very widely separated areas. Within the Laramie group, however, conclusions of this nature are comparatively reliable, and when more is known of this flora and of the characteristic types of different horizons within it, and different areas occupied by it, there can be no doubt that its value in the determination of the precise horizon of new beds both within and without that group must be very great. The following words of Mr. Meek, after a careful survey of the question from the point of view of the invertebrate paleontologist, are equally true for fossil plants: "But it may be asked," he says, "are we to regard all such fossils as of no use whatever in the determination of the ages of strata? Certainly not, because, even in case future discoveries in this country and the Old World should never modify the present conclusions in regard to the geological range of * * * these types * * * so as to enable us to use them with more certainty as a means of drawing parallels on opposite sides of the Atlantic, they will undoubtedly be useful, when viewed in their specific relations, for the identification of strata within more limited areas. That is, when all or most of the details of the stratigraphy of the whole Rocky Mountain region and the vertical range of species have become well known, these fossils will perhaps be found nearly as safe guides in identifying strata at one locality with those of others there, as many other kinds." [1]

But there is a higher ground on which investigations of this nature may be justified. However negative the results may prove, in seeking to make wide generalizations, either for geology or for biology, every new form discovered widens our knowledge of what has been taking place on the surface of the earth since its crust was formed, and the additional knowledge we thus gain of the history of the globe is worth for its own sake all that its laborious pursuit costs, and this quite aside from the added value it possesses in furnishing an ever widening basis for the true laws of both geologic and biologic development.

RECENT COLLECTIONS OF FOSSIL PLANTS FROM THE LARAMIE GROUP.

I have now completed the review of the flora of the Laramie group which, as stated at the outset, would constitute the first part of this memoir, and will now present the concluding portion, also outlined at the beginning, which will be of a somewhat personal character, and will consist of an attempt to record so much of the little that I have been able to contribute to the stock of knowledge relative to the Laramie flora as has thus far assumed a sufficiently definite form. It is, however,

---

[1] Report of the United States Geological Survey of the Territories. F. V. Hayden, Geologist-in-charge. Vol. IX. A Report on the Invertebrate Cretaceous and Tertiary Fossils of the Upper Missouri country. By F. B. Meek, p. lxi.

proper to state that the record I have made will not be complete until I shall have bestowed a large amount of attention and study upon the material in hand. The specimens figured can scarcely be said to have been selected as representative of my collections, although they are so to some extent, but they rather indicate what forms had been sufficiently studied at the time I began to prepare this paper to warrant publishing the figures. The names which I have affixed to them are therefore provisional only, and subject to alteration in the course of the preparation of my final report, which has been merely arrested long enough to enable me to prepare and present in the present synopsis some general considerations which would necessarily be crowded out of the detailed work.

My collections were all made in two seasons, that of 1881 and that of 1883. On the first of these occasions I visited a number of the localities belonging to the lower series situated in Colorado and Wyoming. On the second occasion I visited the valleys of the Lower Yellowstone and Upper Missouri Rivers, and found fossil plants in what are undoubtedly typical Fort Union strata. The itinerary and a general description of the field work of these two seasons have been given in my administrative reports for those years.[1]

COLLECTIONS FROM LOWER LARAMIE STRATA.

The collections made at Golden, Colorado, have not proved particularly rich, and probably very little will be found in them that has not already been reported from that locality. Large palm leaves (*Sabal Campbellii*) and numerous fragments of leaves of Platanus, Ficus, etc., were found in a coarse friable sandstone, either ferruginous and light red, or siliceous and gray or white, in the valley between the Front Range and the basaltic Table Mountain on the east. These strata stand nearly vertical and are in immediate juxtaposition to the productive coal beds on the west. The coal mines themselves are worked in vertical beds which have Cretaceous strata on the west and these coarse sandstones on the east, showing that the direction from east to west represents the descent through the several layers and that the coal veins are at the very base of the Laramie at this place. The strata are conformable, and both the Cretaceous and the Laramie are tilted so as to be approximately vertical. At the base of South Table Mountain the strata are horizontal, and the line dividing the vertical from the horizontal strata could be detected at certain points. A measurement from this line to the base of the coal seam was made at one place and showed 1,700 feet of the upturned edges of Laramie strata. It is probable that we here have the very base of the formation.

The geology of Golden is very complicated, but my observations led

[1] Third Annual Report of the United States Geological Survey, 1881-'82; pp. 26-29. Fifth do., 1883-'84, pp. 55-59.

me to conclude that during the upheaval of the Front Range a break must have occurred along a line near the western base of Table Mountain, forming a crevice through which issued the matter that forms the basaltic cap of these hills. The eastern edge of a broad strip of land lying to the west of this break dropped down until the entire strip of land assumed a vertical position or was tilted somewhat beyond the perpendicular. This brought the Laramie on the east side of the Cretaceous with its upper strata at the extreme eastern, while the coal seam at its base occupied the extreme western side of the displaced rock. The degree of inversion varies slightly at different points and may have been much greater in some places. This will probably account for the discovery at one time of a certain Cretaceous shell (Mactra) *above* a vein of coal in a shaft about 4 miles north of Golden, and about which considerable has been said in discussing the age of the Laramie group. I visited the spot, but found the strata so covered by wash that I was unable to determine their nature.

The collections made at the base of South Table Mountain in a dark and very soft, fine-grained, siliceous-ferruginous sandstone, commonly called tufa, were both more abundant and better preserved than those from the valley, and in them have been found several rare and interesting forms. *Ficus irregularis* was one of the most common, and *Berchemia multinervis* was found. Palms abounded, but only as fragments of narrow portions of leaves. On the surface of the ground, quite well down toward the bottom of the valley, were found numerous fragments of palm wood in the silicified state, as chert, very hard and admitting a high polish. The leaf scars are clearly exhibited, and the vascular bundles and ducts are beautifully shown in cross and longitudinal sections.

At the locality known as Girardot's coal mine, some 5 miles east of Greeley, Colorado, on the open plains, Laramie strata were found containing characteristic mollusks in great abundance, but no plants except the wide-spread *Halymenites major*, which occurred in profusion immediately over the shell beds. Large branching forms were found, as well as forms variously curved and crooked. They seem to be to some extent concretionary, and are composed of iron oxide and sand with a little calcite.

At the mouth of the Saint Vrain, near Platteville, where a day was spent, these forms occurred again in equal abundance and variety. Two species were found here, and perhaps three. Specimens of petrified wood from a large stump, probably coniferous, were collected, but no traces of any other form of plant life were detected. At this point we seem again to have the very base of the Laramie overlying a bluish Cretaceous clay.

The collections from Carbon Station, Wyo., are much more satisfactory than those from the Colorado beds. The station and adjacent track of the Union Pacific Railroad at this point are located in a monoclinal valley running north and south, or at right angles to the railroad. A fault occurs near the station by which the strata on the southwest are lower than those on the northeast. The coal seams on the east and north are close to the surface and sometimes crop out. They pass downward from south to north with a dip of about 15 degrees, reaching across the monoclinal valley through which the railroad runs. On the west and south they grow deeper and have mostly ceased to be worked. The fossil plants, which are very abundant, are always above the coal, and the strata in which they are richest lie five to ten feet from the highest coal seams. Immediately above the coal is a layer of arenaceous limestone, which is generally shaly, but sometimes solid and very hard ("fire clay"). Even in this a few plants occur, but it was nearly impossible to obtain them. The plant beds proper are fine-grained more or less ferruginous and calcareous sandstone shales, quite easily worked, and from them some beautiful specimens of Cissus, Paliurus, and other genera were obtained. These beds are doubtless somewhat higher than those of Black Buttes and Point of Rocks, but they are probably within the limits of the Laramie formation and seem to be the equivalent of the Evanston coal.

The locality denominated Black Buttes always refers to the station of this name on the Union Pacific Railroad, 140 miles west of Carbon Station and in full view of the black rock from which it takes its name. This had been reduced to a mere section house at the time of my visit, and all traffic was by freight trains. It is in the valley of the Bitter Creek, and typical Bitter Creek strata are alone seen. The railroad here runs nearly north and south. The strata dip to the southeast. Opposite the station on the east there are about 100 feet of fucoidal sandstone at the base, above which are two prominent coal seams separated by shales. The coal varies in thickness in both seams and is from three to eight feet thick, the lower seam being perhaps the better in quality. Not more than two feet above the lower coal seam the rocks commence to be plant-bearing. They are reddish on the exposed outer surface, but bluish-gray within, somewhat laminated, and consist of a hard, compact, and very arenaceous limestone. They yield beautifully preserved specimens of leaves, which form the only planes of cleavage.

Above the upper coal the shales are very thin, and their surfaces, where not exposed to the weather, are generally covered with a profusion of very small prints of leaves, stems, culms, fronds, etc., but so fragmentary that little can be done with them. Half a mile north of the station the lower coal seam descends to near the level of the railroad, but the succession of the strata can still be made out. The finest specimens found came from beds a mile or more to the northeast of the station, above a coal mine. The fucoids in the sandstone below the coal at Black Buttes are peculiar and instructive. They seem to consist chiefly of *Halymenites major*, which is often weathered out so as to exhibit good specimens, but more frequently these are incased in concretions which attain huge proportions, sometimes having a diameter of six inches. From the ends of these pod-like bodies short sections of

the typical fucoid, with its verrucose surface, often project. These inflated concretions vary in shape from cylindrical to globular, and when
the projecting fucoid is absent we have the simple spherical concretion
which is familiar to all. By careful selection I succeeded in securing a
good series of these forms, which seem very clearly to point to the fucoidal origin of this class of concretions.

Point of Rocks has become a familiar name to paleontologists since
the discovery there of a thin bed of white sandstone containing very
perfectly preserved specimens of fossil plants that proved, upon examination, to constitute a florula somewhat different from that of any
other locality in the West. This spot was visited and most of the much
discussed forms — *Pistia corrugata, Lemna scutata, Trapa microphylla,
Ficus asarifolia,* etc. — were found, but little was added to the previous
discoveries of others. This locality is a mile or more east of the station,
and is situated quite high up the cliff, which is here steep, and the place
is difficult of access. The lower portion of the cliff at most points near
the railroad consists of white fucoidal sandstone, the fucoids being in
a much less perfect state of preservation than at Black Buttes and more
concretionary. Below the fucoidal sandstone, at one point northwest
of the station, there occurs a bed of light gray or nearly lavender colored
clay containing fragments of ferns and conifers, together with *Pistia
corrugata, Sequoia biformis,* and other species found in the white sandstone stratum last described. It does not seem possible that this stratum can dip sufficiently to the west to bring it to the base of the bluff,
and no evidence of a fault was discovered. The color and fine-grained
character of the rock are similar, but the mineral constitution is very
different in the two beds, so that the question as to their possible stratigraphical identity is still open. If the fucoidal sandstone forms the
base of the Laramie, these clay beds must occupy the summit of the
Cretaceous.

Above the massive white sandstone are several coal seams of good
quality. They vary in thickness and disappear at some points so as
to vary also in number, but about five such seams can usually be seen.
Very few dicotyledonous or phenogamous plants exist in the strata
between the coal beds, although these resemble those at Black Buttes
in all other respects. On the contrary, the fucoids abound throughout
all these strata, including those that overlie the highest coal beds.

At one point, nearly opposite the station to the north, a bed was discovered which contained fine specimens of dicotyledonous and other plants.
This bed is located just above the lowest coal seam, and is about half
way from the base to the summit of the escarpment. The plants seemed,
therefore, to occupy a position very similar to those at Black Buttes,
and they occur in the same hard gray very arenaceous limestone. They
were found only at this one point and in a single layer a foot or more
thick, and rocks a few feet distant in either direction were barren of
them. This florula proved very interesting and yielded a number of

forms not elsewhere found. Among these was the small Ginkgo leaf, which I have called *Ginkgo Laramiensis*.[1] (Plate XXXI, Figure 4.)

Several localities within the Green River group were visited, especially in the vicinity of Green River Station and of Granger, but the description of these will be omitted, and an account given only of localities belonging, with considerable certainty, to the Laramie group as it has been defined. But one other such locality was visited in the year 1881, and respecting the geological position of this there is some doubt. This locality lies very near the boundary line between Wyoming and Utah, some forty miles northwest of Granger, on the divide between the Green and Bear River valleys. The Oregon branch of the Union Pacific Railroad was then in course of construction, and construction trains were running sixteen or eighteen miles out from Granger. The line of the railroad survey was followed from this point, and the plant beds occurred in the ridge through which the tunnel was being excavated. The place was then known as Hodges Pass, and my specimens are so labeled. Fresh-water Tertiary deposits prevailed for the first thirty miles or more, but they were observed to dip perceptibly to the east, and at last disappeared about seven miles east of the divide. They were succeeded here by coal seams, with which they were not conformable, the latter dipping strongly to the northwest. Very heavy beds of coal occur in the vicinity of the pass, and some were reported to have a thickness of sixty feet. The ridge through which the tunnel was being constructed contained fossil plants at nearly all points. The rock consists of a coarse, very arenaceous limestone, or calcareous sandstone, the leaves being either scattered without much stratification through the mass and lying at various angles to one another, often much crumpled or folded, or else in matted layers upon one another in parallel planes, and sometimes so abundant that the rock seems to consist almost wholly of them. In either case it was difficult to obtain perfect specimens. The impressions are very distinct, being of a dark color upon the light matrix, and showing the presence of the silicified leaf-substance. Notwithstanding the coarseness of the material the finer details of nervation are often clearly exhibited. At first sight this flora seemed to be exceedingly monotonous, owing to the prevalence of certain lanceolate or linear willow-shaped forms, but a close study of these reveals considerable variety and the presence of several species and two or three genera. With these, however, occur numerous less abundant forms which lend considerable diversity to the flora of this locality.

There are good reasons for believing that these beds belong to the uppermost series of Laramie strata, and until more is known of them they may be regarded as forming a northern member of the Evanston coal field; the plants, however, differ widely from any found elsewhere.

---

[1] Science, Vol. V, June 19, 1885, p. 496, fig. 7.

542 FLORA OF THE LARAMIE GROUP.

COLLECTIONS FROM THE FORT UNION GROUP.

The several localities from which the principal collections made in the season of 1883 were obtained lie along the Yellowstone River, above and below the town of Glendive, which is situated three miles above old Fort Glendive and on the opposite or right bank of the river, at the point where the Northern Pacific Railroad first enters the valley from the east. Sufficiently precise descriptions of the geographical position of each of these beds were given in my administrative report for that year, and these need not be repeated.

The several beds worked for fossils represent, I am convinced, a number of quite distinct epochs separated far enough in time to have allowed important changes in the vegetation to take place. The localities are not far enough apart geographically to account for the great differences in the different floulas, the extreme distance between the remotest beds not exceeding fifty miles. There were only two of the beds that I was tolerably well satisfied were actually synchronous, and these were among the most remote from each other. These beds are those of Iron Bluff and Burns's Ranch. The plant-bearing stratum at Iron Bluff is situated about fifty feet above the level of the river at low water, while that at Burns's Ranch is at the very water's edge and a few feet above and below. If the beds at Burns's Ranch represent a simple continuation of the strata that appear at Iron Bluff, the dip to the north must be somewhat greater than the natural fall in the river, but the distance is about forty miles. Between Iron Bluff and Glendive, however, there occurs an outcrop of marine Cretaceous strata, containing characteristic Fox Hills shells. This forms an anticlinal of some five or six miles along the right bank of the Yellowstone, and again disappears beneath true Laramie strata some distance above the town. On the side toward Iron Bluff the Cretaceous seems to lie entirely below the railroad cutting at the base of the bluff, but the talus of red blocks of ferruginous baked marl obscured this portion and prevented its study. This is the only outcrop of Cretaceous rocks in the entire district visited by me.

The reasons for regarding the Iron Bluff and Burns's Ranch beds as equivalent are chiefly paleontological. The characteristic plant of the Iron Bluff strata was the large cordate leaf which I have designated as *Cocculus Haydenianus*. This occurs also at Burns's Ranch and has been found only in these two localities. The characteristic plant of the Burns's Ranch locality is *Trapa microphylla*, and this also occurs at Iron Bluff and at no other place in the Fort Union group. The remarkable Cryptogam mentioned above occurs in both beds and several of the celastroid leaves are common to the two localities. None of the forms found at these two localities occur at any of the others. The rock differs greatly in appearance, but this difference is mainly due to the former having been subjected to heat, its carbon driven out, and its iron oxidized, turning it bright red, so that it may be regarded as a ferruginous marl; the other is very calcareous, and may be classed as an argillaceous limestone.

The Iron Bluff stratum yielded a considerable variety of plant forms. Besides the large Cocculus leaves, which were present in great abundance (though, owing to their great size, usually in a fragmentary condition), there occurred an immense quantity of stems of a gigantic Equisetum and of monocotyledonous plants. One of the most striking features of this bed was the occurrence almost everywhere of the stems of certain plants marked all over with very distinct diagonal meshes or cross-lines. These lines consist entirely of deeper colored fine streaks, crossing one another with great regularity at a constant angle. They have the appearance of having wound spirally round the stems in two directions, those of each set being all parallel to one another, and thus forming little rhombs where the systems cross. There is no apparent elevation nor depression, but the fine lines of deeper red are seen in cross-section to penetrate the general surface of light buff, showing that they possess some thickness. The diagonal meshes thus formed vary very much in size, from a millimeter to nearly two centimeters across, and this fineness or coarseness seems to be approximately proportional to the size of the stem on which it occurs. This structure first reminded me of the peculiar cross-lines that occur in the broader stems of certain Monocotyledons, such as Sagittaria, Eriocaulon, etc., and Heer has figured a fossil Sparganium stem exhibiting such a structure. *Caulinites sparganioides* of Lesquereux ("Tertiary Flora," plate xiv, figs. 4 and 10) exhibits something faintly analogous to our plant, and Mr. Lesquereux has sought to explain the occurrence of the cross-lines (p. 100). But the resemblance is too distant to be of any service in the solution of the problem. Certain specimens showing a transition to the normal epidermis, with very fine longitudinal striation, make it next to certain that the parts exhibiting this structure are decorticated, and some evidence exists to prove that the lines may represent the cell walls of the loose cambium tissue of an exogenous plant. The peculiar mode of branching of some specimens also suggests the exogenous rather than the endogenous mode of growth. Certain it is that the diagonal meshes always occur in connection with definite vegetable structure, and even should they prove to be themselves inorganic and to have no connection with the tissues of the plants on which they occur, still the fact must remain that they exist in consequence of such tissues, and are in so far of vegetable origin. I leave the question unsettled for the present and intrust its solution to further research.

The matrix in which the leaf prints found at Burns's Ranch are embedded is an exceedingly fine-grained argillaceous limestone of a bluish-gray color, weathering reddish-brown, and having no regular stratification, but very brittle, and easily breaking at any point with conchoidal fracture, leaving very sharp edges. The degree of friability is much in-

creased by saturation, which was well shown in those fragments that
were taken from below the surface of the water in the river. The surfaces of the leaves often form planes of cleavage, and thus many beautiful specimens were obtained, but the tendency to forsake these planes
and break out at other places rendered many of the specimens fragmentary. Some very perfect specimens of Trapa were obtained. This
plant, as is well known, grows in deep water, from a long submerged
stem, which reaches the surface and bears at its summit a cluster of
small roundish leaves on petioles of different lengths, which are so arranged upon the stem that all the leaves can lie upon the surface of
still water. The longest petioles bear the outer circle of leaves and
successively shorter ones those of circles nearer and nearer the center, where the leaves are small and sessile. Several of my specimens
as well as some of those collected the year previous by Dr. White
show these concentric rosettes of leaves in an interesting way.

The Cocculus leaves are rare in these beds, but several of the best
specimens were nevertheless found here. Numerous fine specimens of
Populus were obtained, only a few of which are figured for this paper.
The sharply serrate, more or less elongated, leaves that seem to belong
to the order Celastrineae were among the most numerous and are nearly
or quite all new to science. A few very fine specimens of the remarkable tapeworm-like Cryptogam mentioned above were found here, but this
form is not yet figured. The bulbous tufted base is much smaller than
in the Iron Bluff specimens, but the remarkable serpent-like rays, with
inflated transversely-ribbed heads and finely-toothed middle portion,
are shown with great clearness.

These two beds (Iron Bluff and Burns's Ranch) appear to me to form
the base of the Fort Union deposit, and present a flora entirely different
from that of any other yet discovered. It is remarkable that the Trapa
found in both of them appears to be the same species as that found so
sparingly in the fine white sandstone layer at Point of Rocks, and what
is still more remarkable, I also found at Burns's Ranch a few specimens
of the characteristic Point of Rocks plant *Pistia corrugata.* I am inclined to regard these two beds as synchronous, and the differences in
the rest of their floras may be accounted for by differences of latitude
and the other conditions previously pointed out. Both seem to occupy
the base of the Laramie and to overlie the same marine Cretaceous deposit.

In ascending the Yellowstone the next locality is that known as
Seven Mile Creek, or Gleason's Ranch. The little stream called Seven
Mile Creek, five or six miles above the mouth of which the ranch is
located, is situated about seven miles below old Fort Glendive, making
it about ten miles below the village of Glendive. Its lower valley is
open and shows no exposures, but at Gleason's Ranch it has narrowed,
and is bounded by hills that rise on the left bank, by a series of terraces,

to a height of about 600 feet. At numerous points along this escarpment good exposures occur, and vegetable remains of one form or another were seen at nearly all elevations. The lowest of the plant beds was not over forty or fifty feet above the valley of the creek, and the plants here consisted almost wholly of the large-leaved Sapindus which is figured on Plate L, Figs. 4–8. A few feet above this occurs a bed of coniferous plants, and immediately above this one yielding a variety of Dicotyledons. Next in order is a stratum of heavy ironstone. This contained a great number of seeds and fruits which are exceedingly curious, but which are as yet wholly undetermined. Mixed with them are leaves in a bad state of preservation belonging to the genus Platanus, and probably to several other genera.

The next bed that proved profitable to work was some 400 feet higher. It was literally filled with leaf impressions, and among these was the immense Platanus leaf, which is here figured natural size, Plate XLI, Fig. 1. Here, too, were found the specimens of Ginkgo, which are also reproduced in our illustrations, and which appear nearly identical with *G. adiantoides* of Unger and quite too near the living plant. Not less interesting was the discovery of the very perfect Sparganium heads, especially those borne on the original stem, one of the specimens of which is shown in the illustrations (Plate XXXII, Fig. 6).

Finally, in the white marl cliff that forms the summit of the series of terraces another florula was found, differing widely from all the rest and characterized by the presence in great abundance of the remarkable leaf which I have called *Credneria daturæfolia* (Plate LVII, Plate LVIII, Figs. 1–5). Associated with this form were many leaves of Populus and Corylus, which were obtained in profusion and in great perfection. This cliff showed evidence of having once been capped by a yellow ferruginous sandstone containing fucoids. One much weatherworn specimen was obtained.

This remarkable series of plant-bearing beds begins at the base with a light-colored and slightly arenaceous limestone, grows less calcareous and more argillaceous and ferruginous until the iron-stone bed is reached. It then presents a series of alternating beds of limestone and ferruginous marl to the Sparganium bed, which is scarcely at all ferruginous. The Credneria cliff consists of a soft, white, and nearly pure marl, slightly tinted on weathered surfaces with iron oxide. The substance of the leaves imbedded in this matrix is clearly visible, and gives the impressions a very dark carbonaceous or lighter brown or lignite colored appearance.

Judging from the slight northerly dip of the strata from the base of the Laramie below Iron Bluff, where it is seen to rest on the Fox Hills, and from Burns's Ranch, where the lowest strata lie beneath the bed of the river, it seems probable that the summit of the Credneria cliff is from 1,200 to 1,500 feet above the base of the Laramie.

The locality on Clear Creek, fifteen miles above Glendive and about three miles back from the river, yielded the largest quantity of fossil plants, but the flora was more uniform than that of other points and consisted chiefly of Viburnum leaves, which seemed when collected to belong almost entirely to one species, but upon closer study they prove to vary considerably and embrace a number of distinct forms. The other kinds of plants, too, which in comparison seemed very few and meager, prove, when separated from the Viburnum leaves and carefully studied, to be quite numerous and varied. Very large and some quite perfect leaves of *Platanus nobilis*, and of the species that possesses the remarkable basal lobe (*P. basilobata*, Plates XLII and XLIII), occurred here, as well as Ulmus leaves, Equisetum tubers, and Leguminosites fruits. In intimate connection with the abundant Viburnum leaves, and not always easy to distinguish from Equisetum and Leguminosites, there were scattered through the shales, always in single detached form, many ovate or elliptical lanceolate fruits, with deep longitudinal furrows (Plate LXII, Figs. 2–6), which, upon careful comparison, I am convinced are the seeds of the Viburnum. This fact would not possess so great importance were it not that certain leaves apparently identical with the most abundant kind found at Clear Creek had been previously collected from the Fort Union group and referred to a different genus. The discovery of these fruits in such immediate relation to the leaves confirms in a very satisfactory manner the conclusion which I had previously reached and expressed that the leaves published by Dr. Newberry as *Tilia antiqua* belonged really to the genus Viburnum.

Most of the plants collected on Clear Creek came from a single stratum about three feet in thickness, which could be traced for long distances along the cliff on the left bank of the creek valley and within from twenty to fifty feet of its summit. The rocks consist of a limestone shale which is so argillaceous as almost to deserve the name of marl, slightly ferruginous, light gray, and very compact. The layers are quite thick, sometimes almost massive, so that very heavy specimens had to be transported; but at some points a true compact marl occurs, which breaks with ease in both directions and has a conchoidal fracture.

Some nine miles farther up the broad valley of Clear Creek occur some elevated ledges, which were visited. On the top of an isolated butte in this locality a bed of compact marl of very friable character was found, containing leaf impressions. This florula was entirely different from that of the locality farther down, and in fact from any other met with on the Yellowstone. The impressions were very clear, but it was difficult to obtain entire leaves, owing to the ease with which the rock would break across the plane of stratification. It was here that were found the very remarkable digitate Aralia-like leaves figured below (Plate XLVIII, Figs. 10–12, Plate XLIX, Fig. 1). Some of the finest specimens of Corylus also came from this bed, and a peculiar fucoid (*Spi-

raxis bivalvis, Plate XXXI, Fig. 3) was abundant, having spiral stria-
tions, as if twisted. This fucoid always exhibited a tendency to split
open longitudinally into two equal valves, and many of the segments
lay around in halves, the plane of division being always smooth and even
and passing directly through the center of the specimen. Only a small
collection was made at this point.

The characteristic fossil of the Cracker Box Creek beds was a species
(or two very closely related species) of Viburnum (*V. asperum*, Newby.,
Plate LXIV, Figs. 4–9, *V. Newberrianum*, Plate LXIV, Figs. 10–12, Plate
LXV, Figs. 1–3), which, however, differs very much from the abundant forms of Clear Creek and does not occur there, nor does the
Clear Creek form occur at Cracker Box Creek, although the two localities are only five miles apart and very similarly situated. On the right
bank of the valley occurred beds containing Populus leaves, masses of
Taxodium Europæum, not elsewhere met with, and an abundance of
both Equisetum and cane (Arundo?), the latter very large. On the
left bank occurred the principal Viburnum bed, and in this a few other
plants were found.

The rock in which the specimens from this locality were embedded is a
highly calcareous marl, sometimes amounting to argillaceous limestone
and slightly ferruginous. At certain points it is of a dark blue color,
sometimes nearly black, and in one fossiliferous bed the outer portion of
a small *butte* which was cut through by a gulch was of a red color, like
that of Iron Bluff, while the interior was blue or dark. This was of
course due to combustion of the carbonaceous matter, the effect of
which had not penetrated to the center of the butte. This combustion
did not affect the character of the plant impressions, but the unburned
portion was much more easily worked and much heavier. In a few of
the oxidized buff specimens from this place, the peculiar diagonal marking, so striking at Iron Bluff, appears. It seems in these cases to occur
on the large gramineous culms.

The several localities on the Yellowstone River above described were
all visited by Dr. C. A. White and his party the year previous, and their
stratigraphical position determined; but, nevertheless, wherever it was
possible I observed and collected the molluscan forms, which, however,
were very rare. The following shells accompany my collections and
have been kindly named for me by Dr. White:

From Iron Bluff: Sphærium (planum?); Physa (Canadensis?).

From Barns's Ranch: Acroloxus minutus.

From Seven Mile Creek: Ironstone bed: Viviparus (species indeterminable); Unio
(species indeterminable); scale of a gar. Sparganium bed: Sphærium (species indeterminable).

From Clear Creek: Physa Canadensis, Whiteaves, ined.: Helix (Patula) (species
undescribed).

From Cracker Box Creek: Viviparus prudentius, White; fragments of gasteropods.

Very few fossil plants were collected during the journey that was

made in August and September down the Missouri River from Fort Benton to Bismarck; but observations that were made upon the Laramie strata as seen at different points, and upon the vegetable remains found in them during that journey, may fittingly be recorded here.

This formation was first met with as the Judith River group, near Birch Creek, about 100 miles below Fort Benton. It here presented the massive sandstone stratum at its base similar to that of the Bitter Creek deposits and appeared about 600 feet above the river, resting upon the Cretaceous. Above this sandstone a few plant remains were found in a soft, whitish-gray marl bed, too imperfect for specific identification, but showing the presence of Equisetum and coniferous and monocotyledonous plants.

Before reaching this point, and much of the way from Coal Banks, an extensive system of dikes of micaceous basalt was observed cutting through the white Cretaceous sandstone in all directions and forming picturesque objects along the river. These seemed to disappear as the Judith River beds came into view, leaving the question of their age relative to that of these beds unsettled; but at a point 18 miles below Claggett a single one of these dikes was observed to rise entirely through the Cretaceous and Laramie strata, both of which were here exposed, thus proving conclusively that the upthrow of lava which produced these dikes occurred posterior to the deposit of at least a large portion of the Judith River strata.

From a point about fifteen miles below Grand Island, where the Judith River group may be said to end, to the Muscle Shell, where the Fort Union group proper may be said to begin, no Laramie strata can be seen, and for much of the distance from the mouth of the Muscle Shell to Poplar Creek, 100 miles above the mouth of the Yellowstone, they merely cap the hills or are wanting altogether. Below Poplar Creek they come down to the level of the river, and some twenty or thirty miles below that point fossil plants were found, including Populus and other Dicotyledons, as well as Conifers, at three different horizons in the cliffs on the right bank of the river. At other points between this and Fort Union, stems of cane and Equisetum were common, but no rich plant beds were found. The Laramie hills here often form nearly perpendicular walls along the south bank of the river and thick beds of coal may be traced for great distances. Much of the Carbonaceous rock has been burned; and at one point the fire was still burning, the rocks in the vicinity of a smoking crevasse being hot, but no actual ignition being visible from without. The progress of this combustion could often be easily traced along a vertical escarpment and the lines clearly seen which were formed by its cessation. At one place the transition from brick red to dark slate color was abrupt along a vertical line extending from top to bottom of a wall several hundred feet high, forming a very striking contrast.

At a point about thirty miles below Fort Buford an interesting bed of northern drift was observed, forming a layer about two feet thick, close down to the water's edge. One hundred miles below Fort Buford a fine deposit of typical Fort Union plants was found, the light slate-colored marl containing them being, however, quite soft. At Little Knife Creek another bed was examined. The Fort Union group is the only deposit in view throughout all this region. Plants were seen at nearly all points that were examined, and at Fort Stevenson I visited a range of low red buttes three miles east of the fort, where I collected a number of good specimens. They closely resembled the forms of the Lower Yellowstone and those previously described from various points within the Fort Union group.

Below this point the country is more flat, the hills are lower and more distant from the river, and there is evidence that the Laramie deposits are passing below the surface. Square Butte, eight or nine miles above Bismarck, is capped by strata that appear to occupy the summit of the formation.

LIST OF SPECIES ILLUSTRATED.

The proportions which this paper has assumed preclude any explanatory remarks upon the figures which I have selected to illustrate the recent collections above described from the Laramie group, and all that can be added in explanation of them is a simple list of the names of the species as they have been decided upon up to this time, leaving more ample discussion of the nice points involved, and the statement of the evidence for or against these determinations, for a subsequent publication. This effort must be regarded as tentative, and subject to much alteration as more thorough study of all the material in hand shall throw additional light upon the many knotty problems involved.

CRYPTOGAMS.

Fucus lignitum, Lx. Plate XXXI, Figs. 1, 2.

Point of Rocks, Wyoming : white sandstone bed east of station (Fig. 1). Burns's Ranch, Montana (Fig. 2).

Spiraxis biralvis, n. sp. Plate XXXI, Fig. 3.

Head of Clear Creek, Montana.

CONIFERÆ.

Ginkgo Laramiensis, Ward, Science, Vol. V, June 19, 1885, p. 496, fig. 7. Plate XXXI, Fig. 4.

Point of Rocks, Wyoming : gray sandstone bed north of station.

Ginkgo adiantoides, Ung. Plate XXXI, Figs. 5, 6.

Seven Mile Creek, Montana : Sparganium bed.

Sequoia biformis, Lx.   Plate XXXI, Figs. 7–12.

Point of Rocks, Wyoming; white sandstone bed east of station (Figs. 7, 8); white marl bed northwest of station (Figs. 9–12).

MONOCOTYLEDONS.

Phragmites Alaskana, Heer.   Plate XXXII, Figs. 1–3.

Burns's Ranch, Montana.

Lemna scutata, Dawson.   Plate XXXII, Figs. 4, 5.

Burns's Ranch, Montana.

Sparganium Stygium, Heer.   Plate XXXII, Figs. 6, 7.

Seven Mile Creek, Montana.

DICOTYLEDONS.

Populus glandulifera, Heer.   Plate XXXIII, Figs. 1–4.   Fig. 3*a*, enlarged.

Burns's Ranch, Montana.

Populus cuneata, Newby.   Plate XXXIII, Figs. 5–11.

Seven Mile Creek, Montana; Sparganium bed (Figs. 5–10). Clear Creek, Montana (Fig. 11).

Populus speciosa, n. sp.   Plate XXXIV, Figs. 1–4.

Clear Creek, Montana.

Populus amblyrhyncha, n. sp.   Plate XXXIV, Figs. 5–9; Plate XXXV, Figs. 1–6.

Seven Mile Creek, Montana ; white marl bed.

Populus daphnogenoides, n. sp.   Plate XXXV, Figs. 7–9.

Seven Mile Creek, Montana ; white marl bed.

Populus oxyrhyncha, n. sp.   Plate XXXV, Figs. 10, 11.

Seven Mile Creek, Montana ; white marl bed.

Populus craspedodroma, n. sp.   Plate XXXVI, Fig. 1.

Burns's Ranch, Montana.

Populus Whitei, n. sp.   Plate XXXVI, Fig. 2.

Burns's Ranch, Montana; collected by Dr. C. A. White in 1882 and named in his honor.

Populus hederoides, n. sp.   Plate XXXVI, Fig. 3.

Seven Mile Creek, Montana ; white marl bed.

Populus Richardsoni, Heer.   Plate XXXVI, Fig. 4.

Burns's Ranch, Montana.

Populus anomala, n. sp.   Plate XXXVI, Fig. 5.

Burns's Ranch, Montana.

Populus Grewiopsis, n. sp.   Plate XXXVI, Fig. 6.

Seven Mile Creek, Montana ; white marl bed.

Populus inæqualis, n. sp.   Plate XXXVI, Fig. 7.

Burns's Ranch, Montana.

Quercus bicornis, n. sp.   Plate XXXVI, Fig. 8.

Seven Mile Creek. Montana ; bed below the ironstone.

Quercus Doljensis, Pilar.   Plate XXXVI, Figs. 9, 10.

Black Buttes Station, Wyoming.

Quercus Carbonensis, n. sp.   Plate XXXVII, Fig. 1.

Carbon Station, Wyoming.

Quercus Dentoni, Lx.   Plate XXXVII, Fig. 2.

Point of Rocks, Wyoming ; gray sandstone bed north of station.

Dryophyllum aquamarum, n. sp.   Plate XXXVII, Figs 3–5.

Black Buttes Station. Wyoming.

Dryophyllum Bruneri, n. sp.   Plate XXXVII, Figs. 6–9.

Point of Rocks, Wyoming ; gray sandstone bed (Figs. 6, 7).   Hodges Pass, Wyoming (Figs. 8, 9).   Named in honor of Prof. Lawrence Bruner.[1]

Dryophyllum falcatum, n. sp.   Plate XXXVII, Fig. 10.

Hodges Pass, Wyoming.

Dryophyllum basidentatum. n. sp.   Plate XXXVII, Fig. 11.

Carbon Station, Wyoming.

Corylus Americana, Walt.   Plate XXXVIII, Figs. 1–5.

Seven Mile Creek, Montana ; white marl bed.

Corylus rostrata, Ait.   Plate XXXIX, Figs. 1–4.

Seven Mile Creek, Montana : white marl bed.

Corylus Fosteri, n. sp.   Plate XXXIX, Figs. 5, 6.

Head of Clear Creek. Montana (Fig. 5) ; Clear Creek, Montana (Fig. 6) ; the latter collected in 1882 by Dr. White's party : the first by Mr. Richard Foster, for whom it is named.

? *Corylus McQuarrii*, Heer.   Plate XXXIX, Fig. 7.

Seven Mile Creek, Montana : bed below the ironstone.

Alnus Grewiopsis, n. sp.   Plate XXXIX, Fig. 8.

Hodges Pass, Wyoming.

Betula prisca, Ett.   Plate XL, Fig. 1.

Seven Mile Creek, Montana ; bed below the ironstone.

Betula coryloides, n. sp.   Plate XL, Fig. 2.

Seven Mile Creek, Montana : white marl bed.

Betula basiserrata, n. sp.   Plate XL, Fig. 3.

Seven Mile Creek, Montana ; white marl bed.

Myrica Torreyi, Lx.   Plate XL, Fig. 4.

Black Buttes Station, Wyoming.

? *Juglans Ungeri*, Heer.   Plate XL, Fig. 5.

Burns's Ranch, Montana.

_____

[1] Professor Bruner's valuable services on this expedition are otherwise acknowledged in my administrative report for that year.   (See Third Annual Report United States Geological Survey, 1881–'82, p. 29).

Juglans nigella, Heer. Plate XL, Fig. 6.

Burns's Ranch, Montana.

Carya antiquorum, Newby. Plate XL, Fig. 7.

Carbon Station, Wyoming.

Platanus Heerii, Lx. Plate XL, Figs. 8, 9.

Black Buttes Station, Wyoming.

Platanus nobilis, Newby. Plate XLI, Fig. 1.

Seven Mile Creek, Montana; Sparganium bed.

Platanus basilobata, n. sp. Plate XLII. Figs. 1–4. Fig. 4*a*, enlarged.
Plate XLIII, Fig. 1.

Seven Mile Creek, Montana; Sparganium bed (Plate XLII). Clear Creek, Montana (Plate XLIII).

Platanus Guillelmæ, Göpp. Plate XLIV, Fig. 1.

Burns's Ranch, Montana.

Platanus Raynoldsii, Newby. Plate XLIV, Figs. 2, 3.

Clear Creek, Montana; collected in 1882 by Dr. White's party.

Ficus irregularis, Lx. Plate XLIV, Figs. 4, 5.

Golden, Colorado.

Ficus spectabilis, Lx. Plate XLIV, Fig. 6.

Golden, Colorado; collected in November, 1881, by Mr. C. W. Cross for Mr. S. F. Emmons.

Ficus Crossii, n. sp. Plate XLIV, Fig. 7.

Golden, Colorado; collected in 1881 by Mr. C. W. Cross for Mr. S. F. Emmons.

Ficus speciosissima, n. sp. Plate XLV, Fig. 1.

Point of Rocks, Wyoming; gray sandstone bed north of station.

Ficus tiliæfolia, Heer. Plate XLV, Fig. 2.

Burns's Ranch, Wyoming.

Ficus sinuosa, n. sp. Plate XLV, Fig. 3.

Black Buttes Station, Wyoming.

Ficus limpida, n. sp. Plate XLV, Fig. 4.

Clear Creek, Montana.

Ficus rhamnifolia, n. sp. Plate XLV, Figs. 5–9.

Clear Creek, Montana.

Ulmus planeroides, n. sp. Plate XLVI, Figs. 1, 2.

Clear Creek, Montana.

Ulmus minima, n. sp. Plate XLVI, Figs. 3, 4.

Clear Creek, Montana.

Ulmus rhamnifolia, n. sp. Plate XLVI, Fig. 5.

Clear Creek, Montana.

Ulmus orbicularis, n. sp.   Plate XLVI, Fig. 6.
Clear Creek, Montana.

Laurus resurgens, Sap.   Plate XLVI, Fig. 7.
Bull Mountains, Montana; collected by Dr. A. C. Peale in 1886.

Laurus primigenia, Ung.   Plate XLVI, Figs. 8–10.
Carbon Station, Wyoming (Fig. 8). Point of Rocks, Wyoming; white sandstone bed east of station (Figs. 9, 10).

Litsæa Carbonensis, n. sp.   Plate XLVI, Fig. 11.
Carbon Station, Wyoming.

Cinnamomum lanceolatum, Heer.   Plate XLVI, Fig. 12.
Hodges Pass, Wyoming.

Cinnamomum affine, Lx.   Plate XLVII, Figs. 1–3.
Black Buttes Station, Wyoming.

Daphnogene elegans, Wat.   Plate XLVII, Fig. 4.
Black Buttes Station, Wyoming.

? Monimiopsis amboræfolia, Sap.   Plate XLVII, Fig. 5.
Seven Mile Creek, Montana; Sapindus bed.

? Monimiopsis fraterna, Sap.   Plate XLVII, Fig. 6.
Seven Mile Creek, Montana; bed below the ironstone.

Nyssa Buddiana, n. sp.   Plate XLVII, Fig. 7.
Hodges Pass, Wyoming. Named in honor of Mr. J. Budd, superintendent of construction of the Oregon branch of the Union Pacific Railroad, who directed me to this locality.

Cornus Fosteri, n. sp.   Plate XLVII, Fig. 8.
Upper Seven Mile Creek, ten miles above Glendive, Montana; collected by Mr. Richard Foster, of Dr. White's party, in 1882.

Cornus Studeri, Heer.   Plate XLVIII, Fig. 1.
Point of Rocks, Wyoming; gray sandstone bed north of station.

Cornus Emmonsii, n. sp.   Plate XLVIII, Figs. 2, 3.
Golden, Colorado (Fig. 2); collected by Mr. S. F. Emmons, in July, 1882. Point of Rocks, Wyoming; gray sandstone bed north of station (Fig. 3).

Hedera parvula, n. sp.   Plate XLVIII, Fig. 4.
Clear Creek, Montana.

Hedera minima, n. sp.   Plate XLVIII, Fig. 5.
Head of Clear Creek, Montana.

Hedera Bruneri, n. sp.   Plate XLVIII, Fig. 6.
Black Buttes Station, Wyoming.

Hedera aquamara, n. sp.   Plate XLVIII, Fig. 7.
Black Buttes Station, Wyoming.

Aralia notata, Lx.   Plate XLVIII, Fig. 8.
Clear Creek, Montana.

Aralia Looziana, Sap. & Mar.   Plate XLVIII, Fig. 9.
Clear Creek, Montana.

Aralia digitata, n. sp.   Plate XLVIII, Figs. 10–12 ; Plate XLIX, Fig. 1.
Head of Clear Creek, Montana.

Trapa microphylla, Lx.   Plate XLIX, Figs. 2–5.
Burns's Ranch, Wyoming.

Hamamelites fothergilloides, Sap.   Plate XLIX, Fig. 6.
Seven Mile Creek, Montana; bed below the ironstone.

Leguminosites arachioides, Lx.   Plate XLIX, Fig. 7.
Clear Creek, Montana.

Acer trilobatum tricuspidatum, Heer.   Plate XLIX, Figs. 8, 9.
Clear Creek, Montana (Fig. 8); collected by Dr. White's party in 1882. Little
Missouri River, Dakota (Fig. 9); collected by Hayden and Peale in 1883.

Acer indivisum, Web.   Plate L, Fig. 1.
Carbon Station, Wyoming.

Sapindus affinis, Newby.   Plate L, Figs. 2, 3.
Gladstone, Dakota; collected by Hayden and Peale in 1883.

Sapindus grandifoliolus, n. sp.   Plate L, Figs. 4–8.
Seven Mile Creek, Montana; Sapindus bed.

Sapindus alatus, n. sp.   Plate L, Figs. 9, 10.
Seven Mile Creek, Montana ; Sapindus bed.

Sapindus angustifolius, Lx.   Plate LI, Figs. 1–3.
Seven Mile Creek, Montana ; Sapindus bed.

Vitis Braueri, n. sp.   Plate LI, Figs. 4, 5.
Carbon Station, Wyoming.

Vitis Carbonensis, n. sp.   Plate LI, Fig. 6.
Carbon Station, Wyoming.

Vitis Xantholithensis, n. sp.   Plate LI, Figs. 7, 8.
Burns's Ranch, Montana.

Vitis cuspidata, n. sp.   Plate LI, Figs. 9–11.
Burns's Ranch, Montana.

Berchemia multinervis, Al. Br.   Plate LI, Figs. 12, 13.
Golden, Colorado.

Zizyphus serrulata, n. sp.   Plate LI, Figs. 14, 15.
Burns's Ranch, Montana.

Zizyphus Meekii, Lx.   Plate LII, Figs. 1, 2.
Carbon Station, Wyoming (Fig. 1). Bozeman Coal Mines, Montana (Fig. 2);
collected by Hayden and Peale in 1883.

Zizyphus cinnamomoides, Lx.   Plate LII, Fig. 3.
Seven Mile Creek, Montana; white marl bed.

Paliurus Colombi, Heer.   Plate LII, Figs. 4–6.
  Burns's Ranch, Montana (Figs. 4, 5). Carbon Station, Wyoming (Fig. 6).

Paliurus pulcherrima, n. sp.   Plate LII, Fig. 7.
  Carbon Station, Wyoming.

Paliurus Pealei, n. sp.   Plate LII, Figs. 8–10.
  Little Missouri River. Dakota; collected by Dr. A. C. Peale in 1883.

Celastrus ferrugineus, n. sp.   Plate LII, Figs. 11–14.
  Burns's Ranch, Montana (Fig. 11); Iron Bluff, Montana (Figs. 12–14).

Celastrus Taurinensis, n. sp.   Plate LII, Figs. 15, 16.
  Bull Mountains, Montana (Figs. 15); Burns's Ranch, Montana (Fig. 16).

Celastrus alnifolius, n. sp.   Plate LIII, Figs. 1, 2.
  Burns's Ranch, Montana.

Celastrus pterospermoides, n. sp.   Plate LIII, Figs. 3–6.
  Burns's Ranch, Montana.

Celastrus ovatus, n. sp.   Plate LIII, Fig. 7.
  Iron Bluff, Montana.

Celastrus grewiopsis, n. sp.   Plate LIII, Fig. 8.
  Burns's Ranch, Montana.

Celastrus currinervis, n. sp.   Plate LIII, Figs. 9, 10.
  Burns's Ranch, Montana.

Euonymus Xantholithensis, n. sp.   Plate LIV, Figs. 1, 2.
  Burns's Ranch, Montana.

Elæodendron serrulatum, n. sp.   Plate LIV, Figs. 3–5.
  Burns's Ranch, Montana (Figs. 3, 4). Seven Mile Creek, Montana (Fig. 5).

Elæodendron polymorphum, n. sp.   Plate LIV, Figs. 6–12.
  Burns's Ranch, Montana.

Grewia crenata (Ung.) Heer.   Plate LIV, Fig. 13.
  Bull Mountains, Montana; collected by Hayden and Peale in 1883.

Grewia celastroides, n. sp.   Plate LIV, Fig. 14.
  Iron Bluff, Montana.

Grewia Pealei, n. sp.   Plate LV, Figs. 1–3.
  Bull Mountains, Montana; collected by Dr. A. C. Peale in 1883.

Grewia oborata, Heer.   Plate LV, Figs. 4, 5.
  Seven Mile Creek, Montana; white marl bed.

Grewiopsis platanifolia, n. sp.   Plate LV, Fig. 6.
  Seven Mile Creek, Montana; Sparganium bed.

Grewiopsis viburnifolia, n. sp.   Plate LV, Fig. 7.
  Burns's Ranch, Montana.

Grewiopsis populifolia, n. sp. Plate LV, Figs. 8-10.
  Burns's Ranch, Montana.

Grewiopsis ficifolia, n. sp. Plate LVI, Figs. 1, 2.
  Black Buttes Station, Wyoming.

Grewiopsis paliurifolia, n. sp. Plate LVI, Fig. 3.
  Black Buttes Station, Wyoming.

Pterospermites cordatus, n. sp. Plate LVI, Fig. 4.
  Seven Mile Creek, Montana; bed below the ironstone.

Pterospermites Whitei, n. sp. Plate LVI, Figs. 5, 6.
  Burns's Ranch, Montana: collected by Dr. C. A. White in 1882.

Pterospermites minor, n. sp. Plate LVI, Figs. 7-9.
  Burns's Ranch, Montana.

Credneria ? daturæfolia, n. sp. Plate LVII, Figs. 1-5; Plate LVIII,
  Figs. 1-5.
  Seven Mile Creek, Montana ; white marl bed
  Plate LVIII, Fig. 6, represents a leaf of Datura Stramonium, L., introduced
  to illustrate the similarity of its nervation to that of the fossil leaves.

Cocculus Haydenianus, n. sp. Plate LIX, Figs. 1-5.
  Burns's Ranch, Montana (Figs. 1-4). Iron Bluff, Montana (Fig. 5).
  Named in honor of Ensign Everett Hayden, U. S. N., who has taken a special
  interest in this plant.

Liriodendron Laramiense, n. sp. Plate LX, Fig. 1.
  Point of Rocks Station, Wyoming; gray sandstone bed north of station.

Magnolia pulchra, n. sp. Plate LX, Figs. 2, 3.
  Point of Rocks Station, Wyoming: gray sandstone bed north of station.

Diospyros brachysepala, Al. Br. Plate LX, Figs. 4, 5.
  Burns's Ranch, Montana (Fig. 4). Seven Mile Creek, Montana (Fig. 5).

Diospyros ficoidea, Lx. Plate LX, Figs. 6, 7.
  Burns's Ranch, Montana (Fig. 6). Clear Creek, Montana (Fig. 7).

Diospyros ? obtusata, n. sp. Plate LX, Fig. 8.
  Seven Mile Creek, Montana; bed below the ironstone.

Viburnum tilioides (*Tilia antiqua*, Newby.). Plate LXI, Figs. 1-7;
  Plate LXII, Figs. 1-6.
  Clear Creek, Montana.

Viburnum perfectum, n. sp. Plate LXII, Figs. 7-9.
  Clear Creek, Montana.

Viburnum macrodontum, n. sp. Plate LXII, Fig. 10.
  Clear Creek, Montana.

Viburnum limpidum, n. sp. Plate LXIII, Figs. 1-4.
  Clear Creek, Montana.

Viburnum Whymperi. Heer. Plate LXIII, Fig. 5.
Clear Creek, Montana.

Viburnum perplexum, n. sp. Plate LXIII, Figs. 6, 7
Burns's Ranch, Montana; collected by Dr. White's party in 1882.

Viburnum elongatum, n. sp. Plate LXIII, Figs. 8, 9.
Clear Creek, Montana.

Viburnum oppositinerve, n. sp. Plate LXIV, Figs. 1, 2.
Clear Creek, Montana.

Viburnum erectum, n. sp. Plate LXIV, Fig. 3.
Clear Creek, Montana.

Viburnum asperum, Newby. Plate LXIV, Figs. 4-9.
Cracker Box Creek, Montana (Figs. 4-8). Seven Mile Creek, Montana ; Sparganium bed (Fig. 9).

Viburnum Newberrianum, n. sp. Plate LXIV, Figs. 10-12; Plate LXV, Figs. 1-3.
Cracker Box Creek, Montana.

Viburnum Nordenskjöldi, Heer. Plate LXV, Figs. 4-6.
Clear Creek, Montana (Fig. 4). Little Missouri River, Dakota (Fig. 6). Gladstone, Dakota. (Fig. 5). The last two were collected by Dr. A. C. Peale in 1883.

Viburnum betulæfolium, n. sp. Plate LXV, Fig 7.
Burns's Ranch, Montana ; collected by Dr. White's party in 1882.

Viburnum finale, n. sp. Plate LXV, Fig 8.
Iron Bluff, Montana.

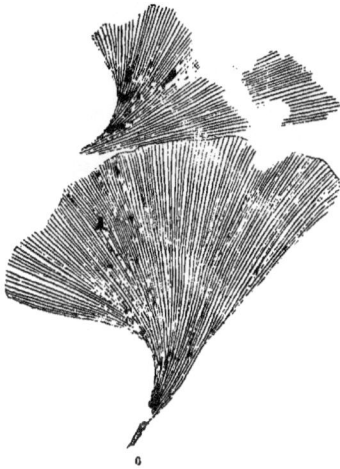

6

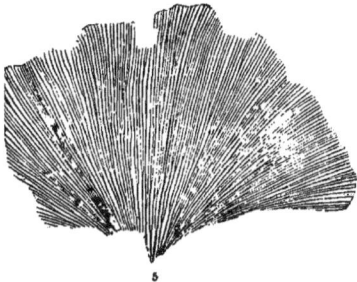

5

10

11

12

Fig 3. Spiraxis bivalvis, n. sp.

'ERÆ.

adiantoides. Ung   Figs. 7-12. Sequoia biformis, Lx.

32

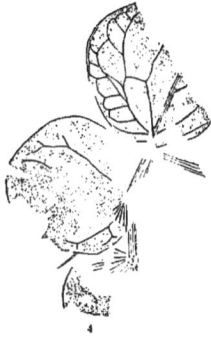

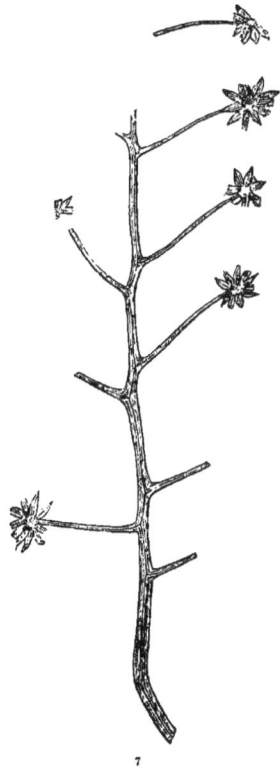

LEDONS.

ontata, Dawson.   FIGS. 6, 7. Sparganium Stygium, Heer.

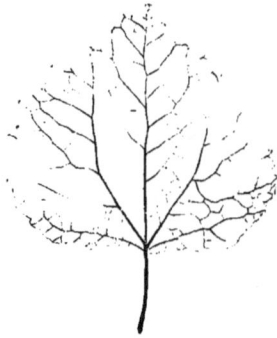

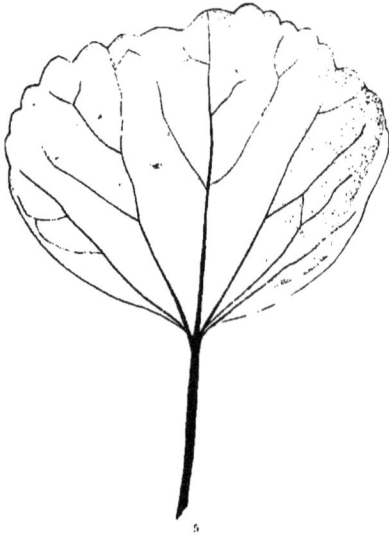

Figs. 1-4 Populus glandulifera, Heer.   Fig. 3a

Figs. 5–11. P. cuneata, Newby.

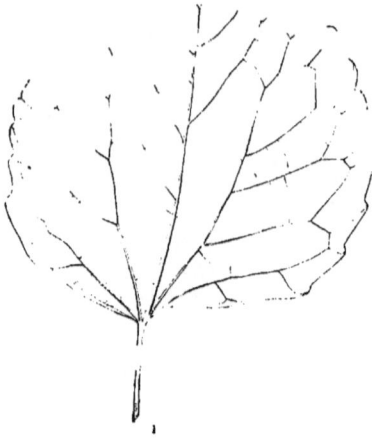

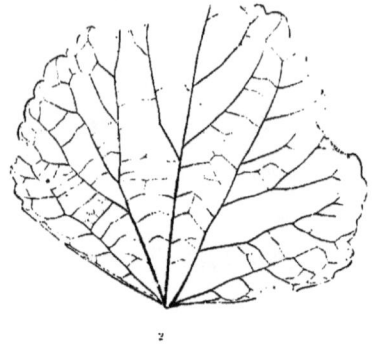

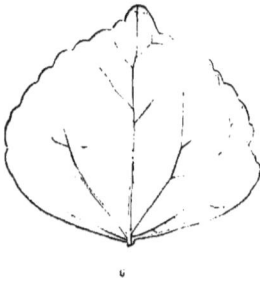

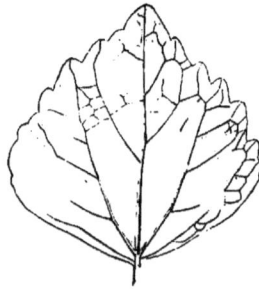

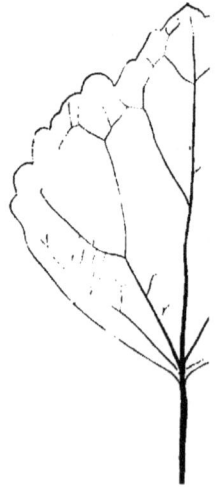

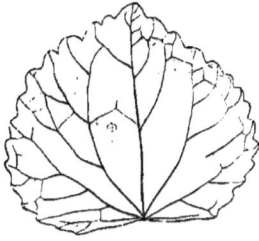

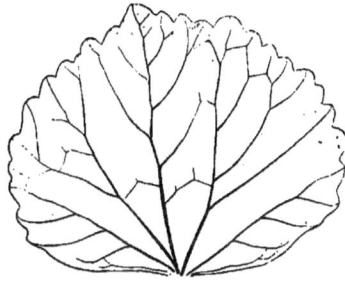

3

4

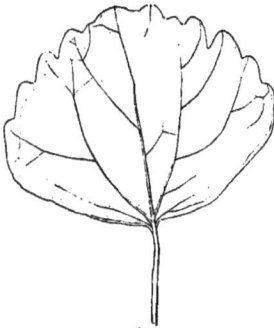

8

9

Gs. 5-9 P. amblyrhyncha. n. sp.

FIGS. 1-6. Populus amblyrhyncha, n. sp.   FIGS. 7-9. P.

hnogenoides, n. sp.   Figs. 10, 11. P. oxyrhyncha. n. sp.

FIG. 1. Populus craspedodroma, n. sp.   FIG. 2. P. Whitei, n. sp.   FIG. 3. P. hederoides, n. sp.   FIG. 4. P. Richardsoni, Heer.   FIG. 5.

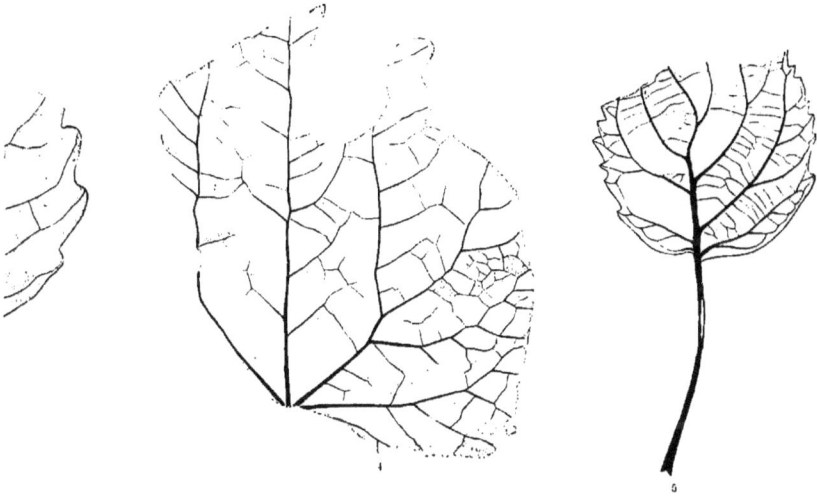

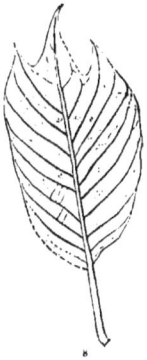

mala, n. sp.   FIG. 6. P. Grewiopsis, n. sp.   FIG. 7. P. inæqualis, n. sp.   FIG. 8. Quercus bicornis, n. sp.   FIGS. 9, 10. Q. Doljensis, Pilar.

37

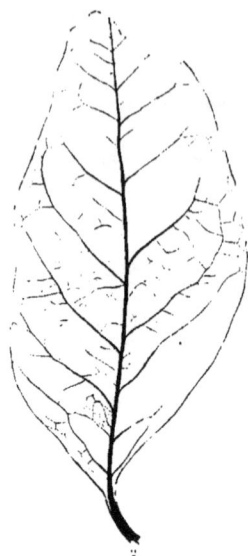

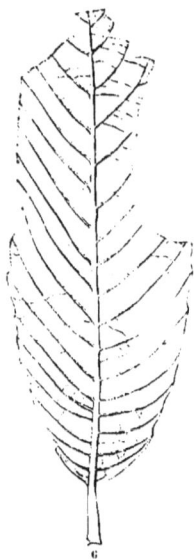

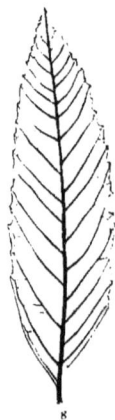

Fig. 1. Quercus carbonensis, n. sp.   Fig. 2. Q. Deutoni, Lx.   Figs. 3-5. Dryophyllum aquam

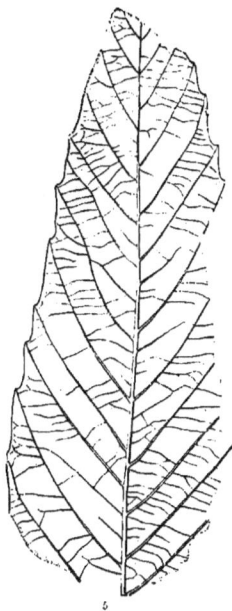

FIG. 10. D. falcatum, n. sp.   FIG. 11. D. basidentatum, n. sp.

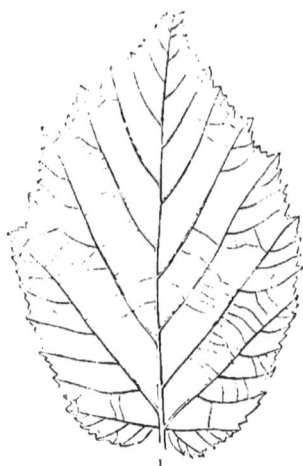

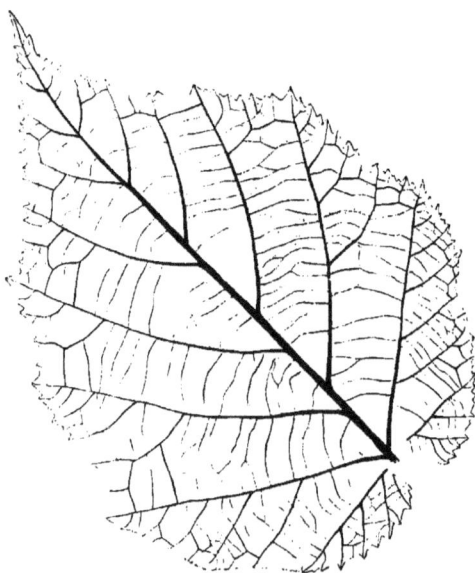

3

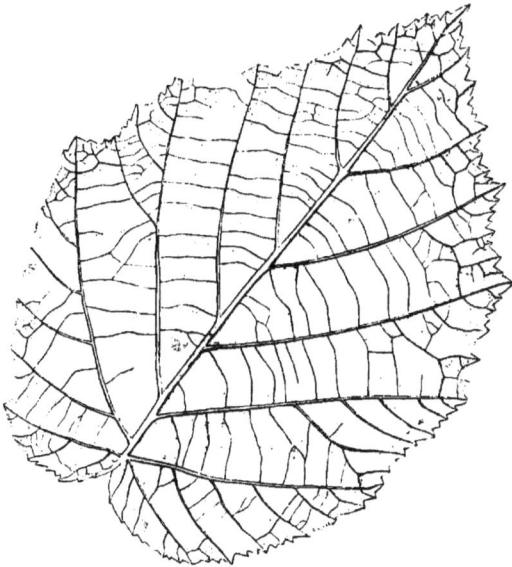

5

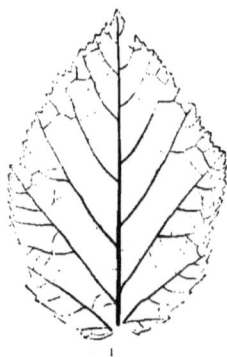

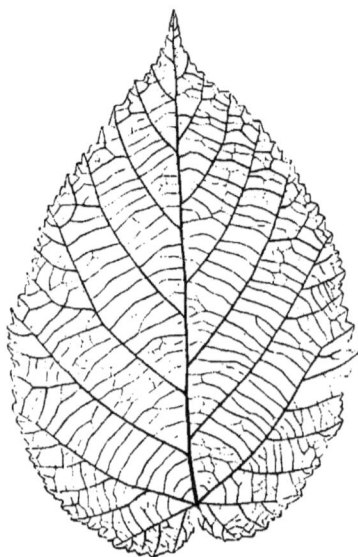

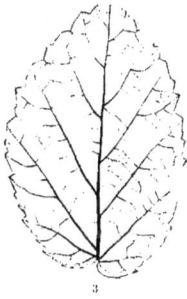

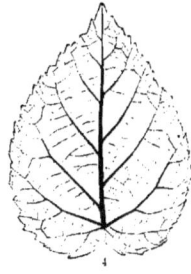

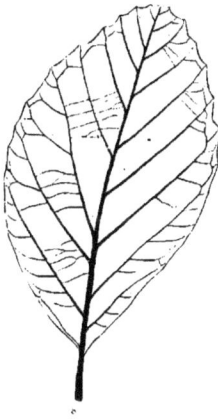

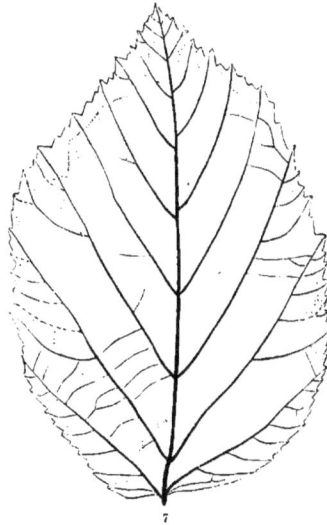

FIGS. 7. !C. McQuarrii. Heer.   FIG. 8. Alnus Grewiopsis. n. sp.

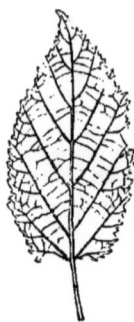

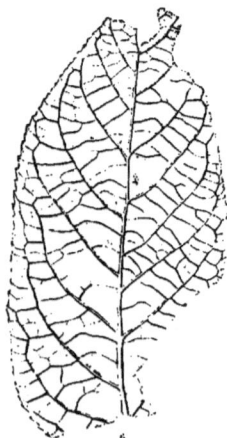

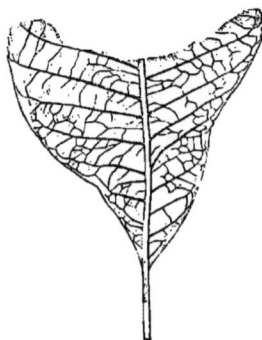

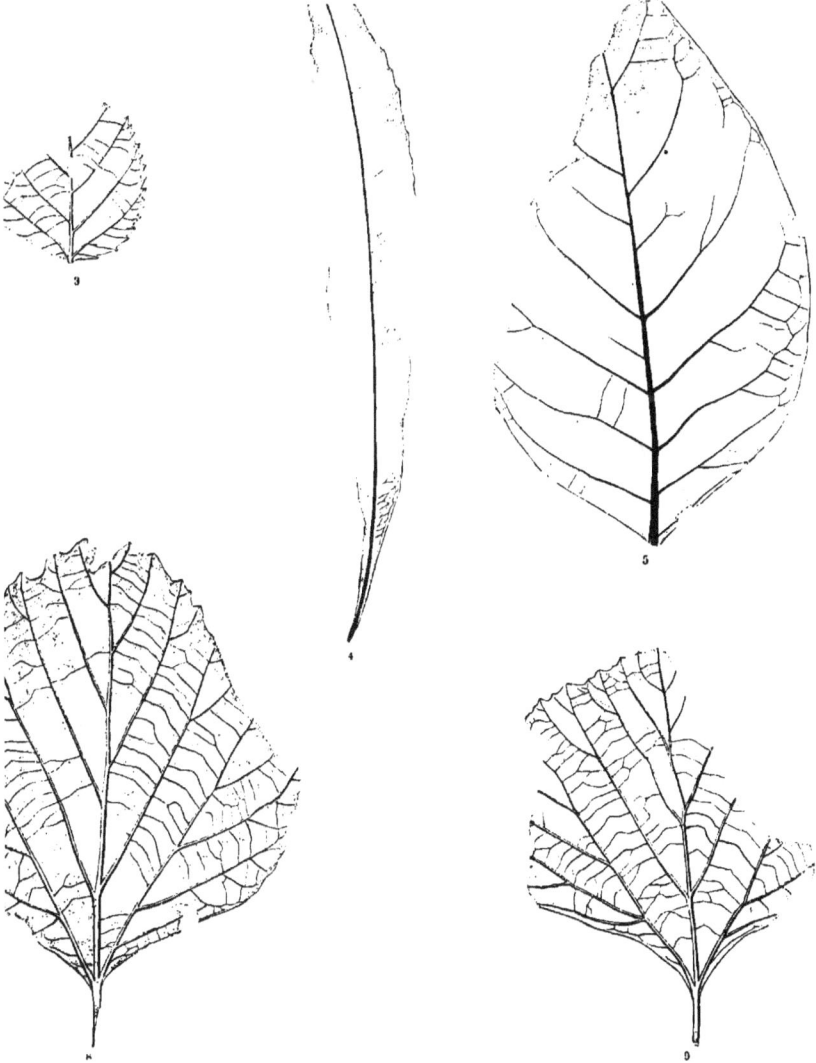

IG. 5. Juglans Ungeri, Heer.   FIG. 6. J. nigella. Ung.   FIG. 7. Carya antiquorum. Lx.   FIGS. 8 9. Platanus Heerii. Lx.

DICOTYLEDONS

Fig. 1. Platanus nobilis. Newby.

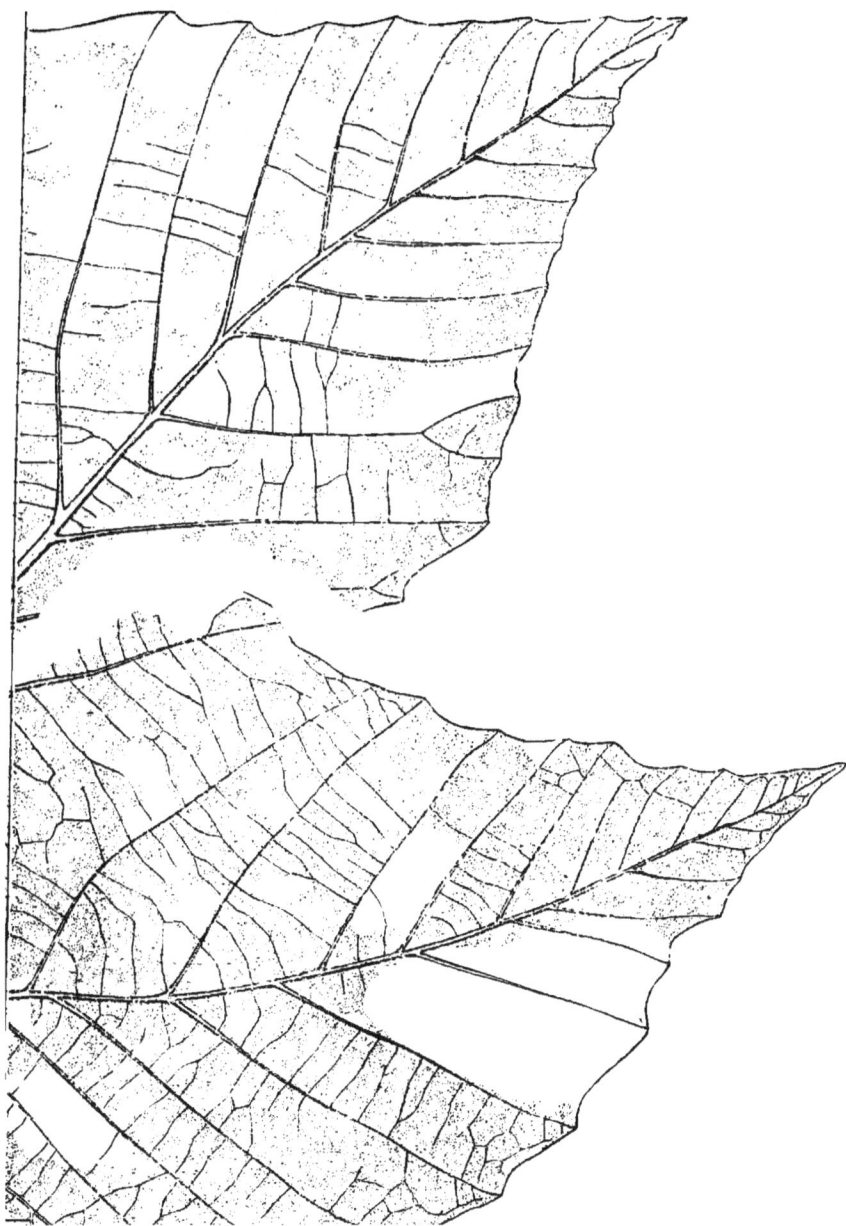

42

FIGS. 1-4. Platauna basilobata

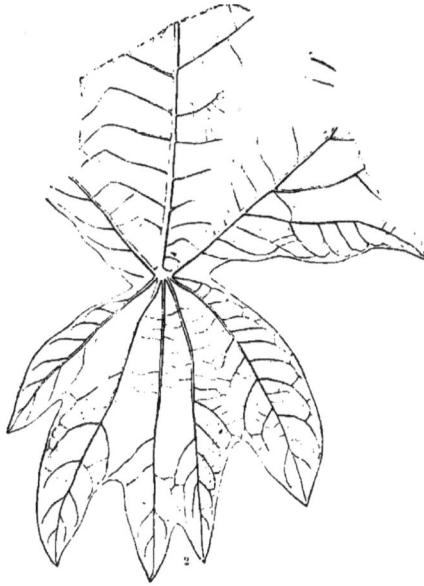

S

Fig. 4a. Enlarged detail.

43

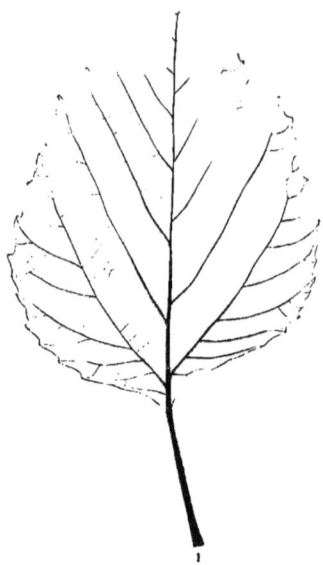

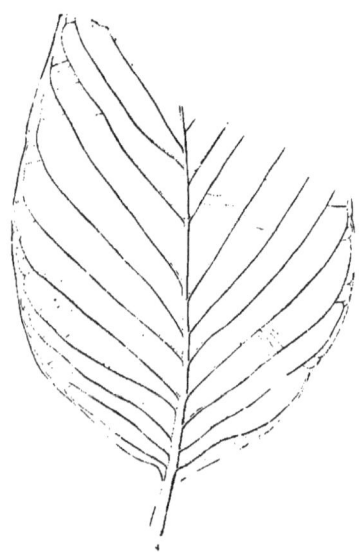

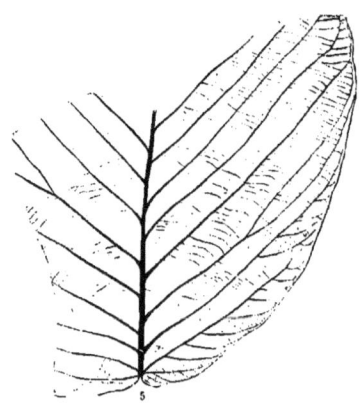

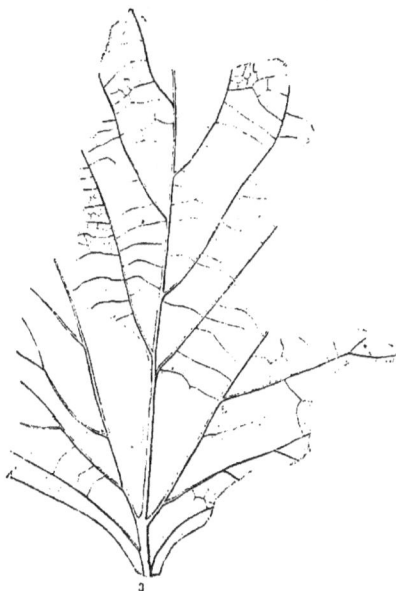

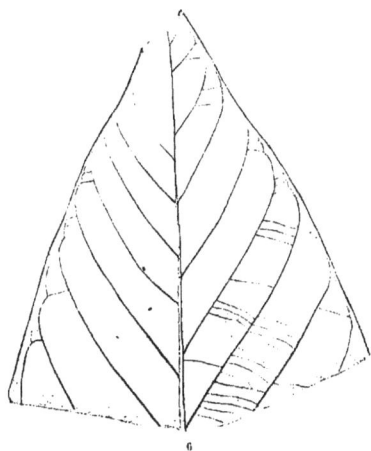

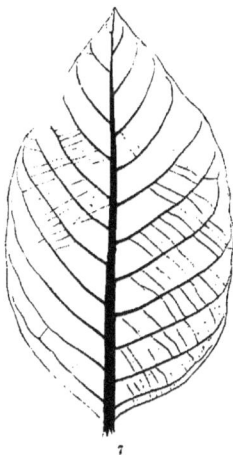

44

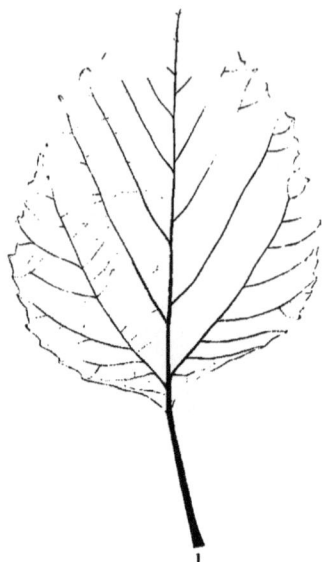

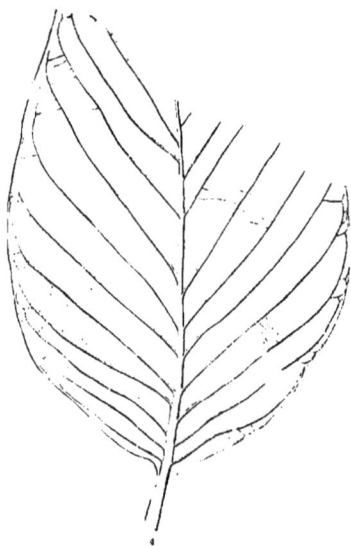

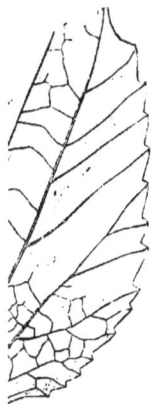

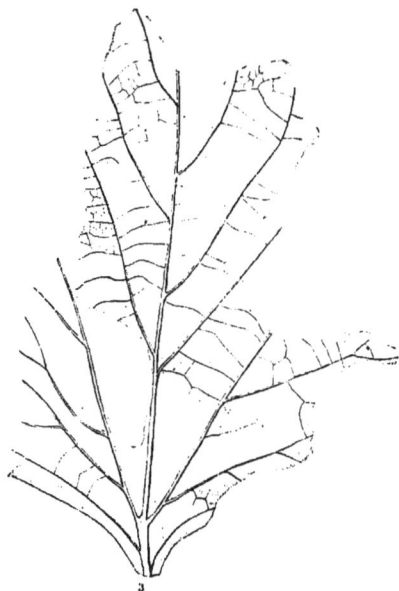

3

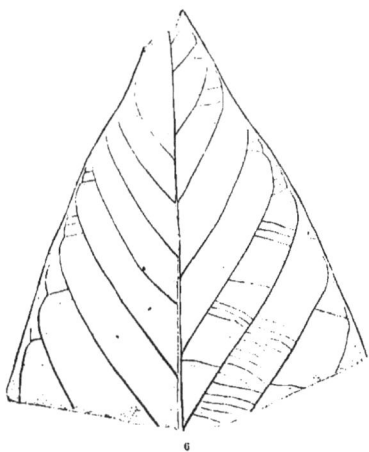

6

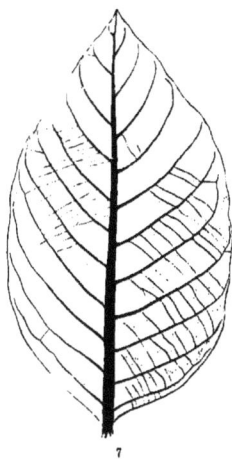

7

45

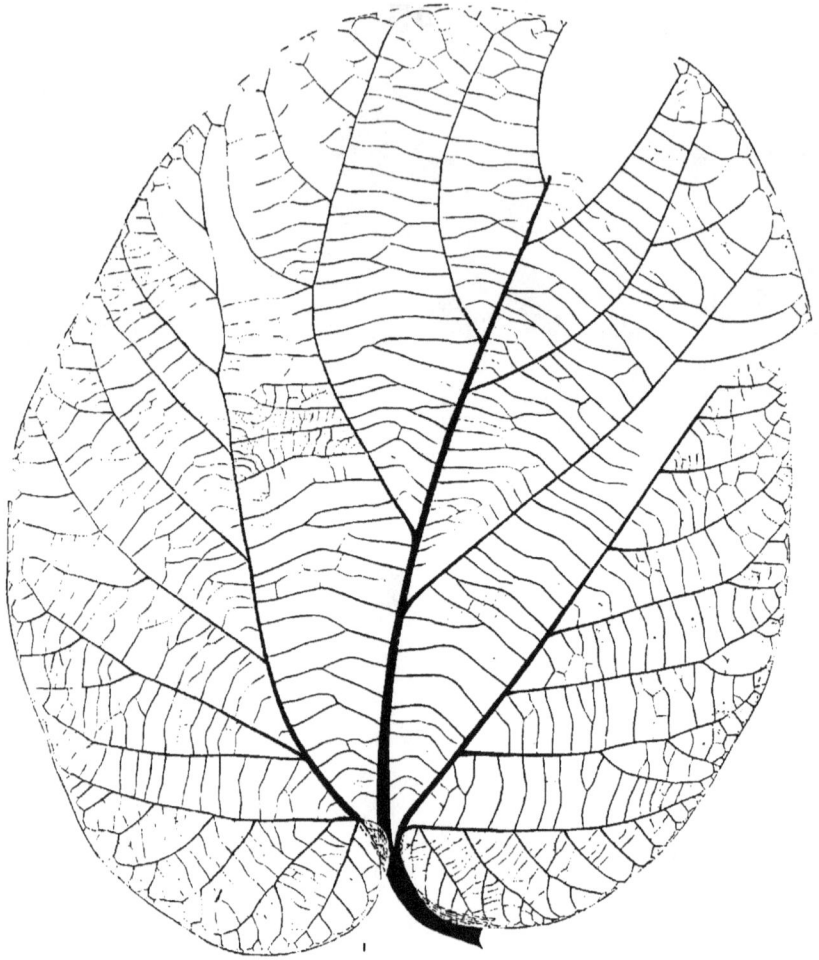

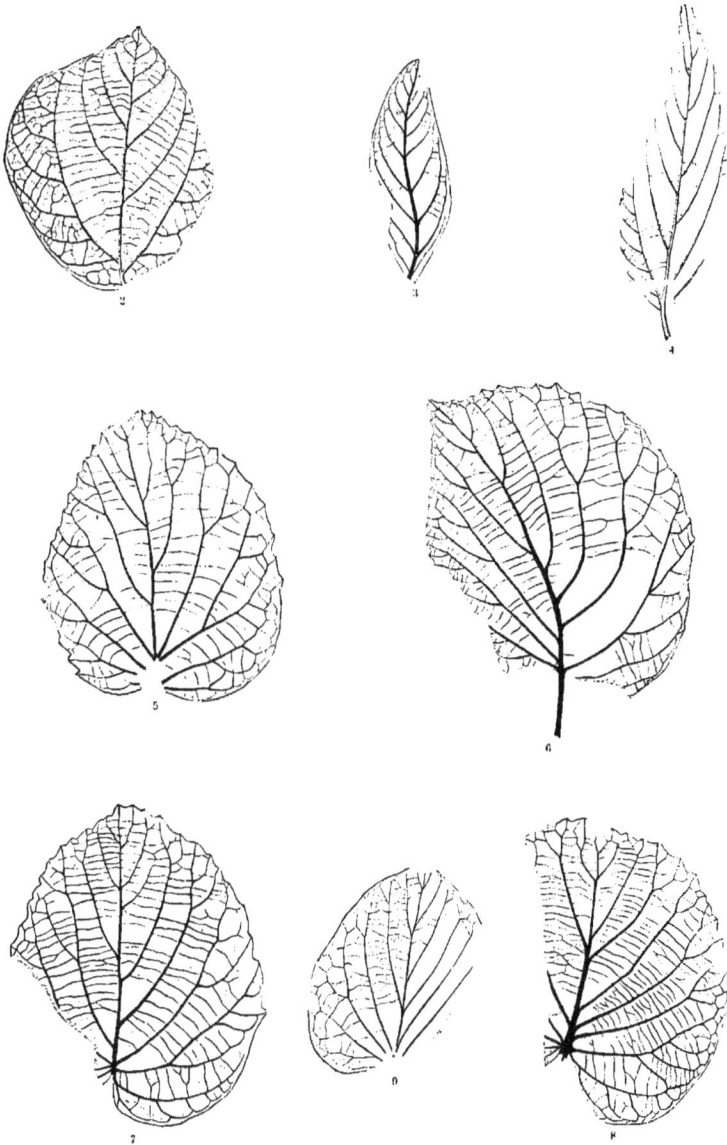

Fig. 4 F. limpida, n. sp.   Figs. 5-9. F. viburnifolia, n. sp.

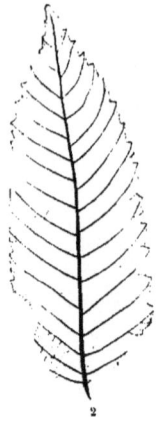

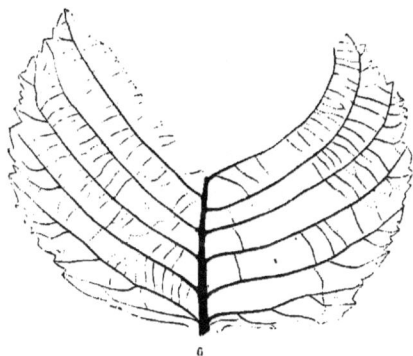

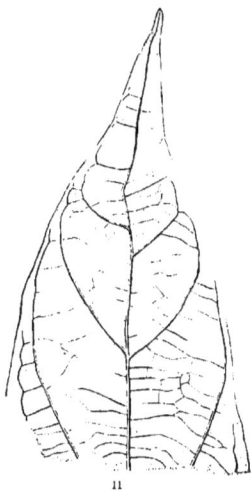

Figs. 8-10. L. primigenia, Ung.   Fig. 11. Litsea Carbonensis, n. sp.   Fig. 12. Cinnamomum lanceolatum. Heer.

47

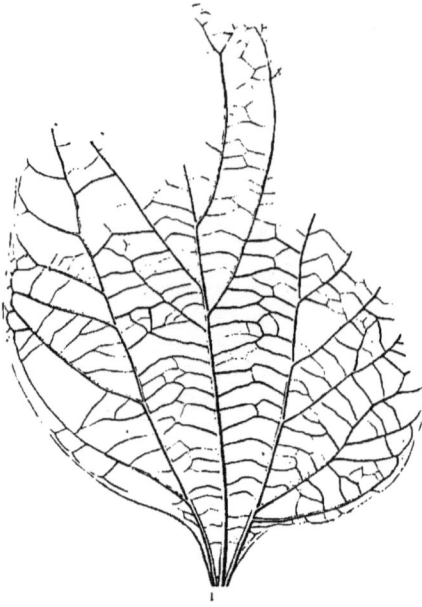

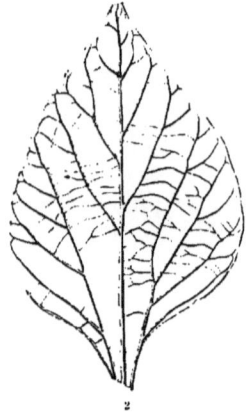

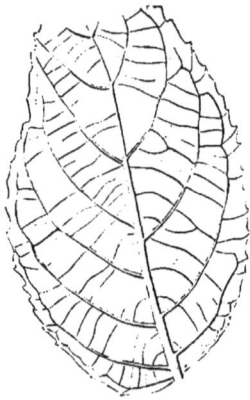

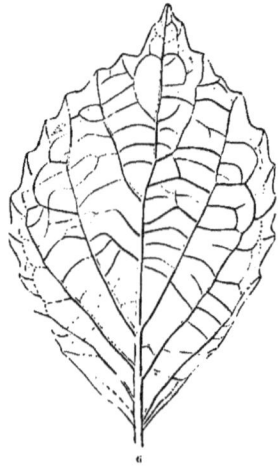

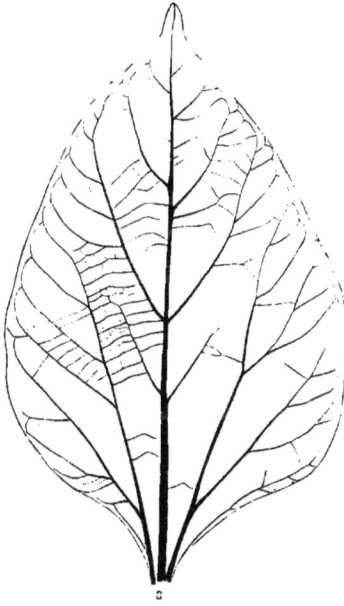

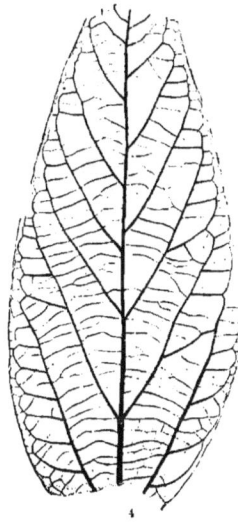

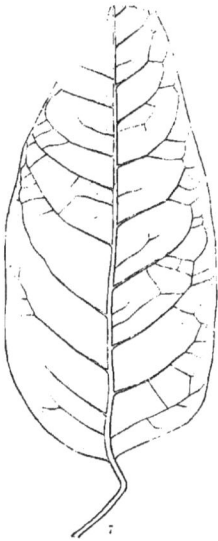

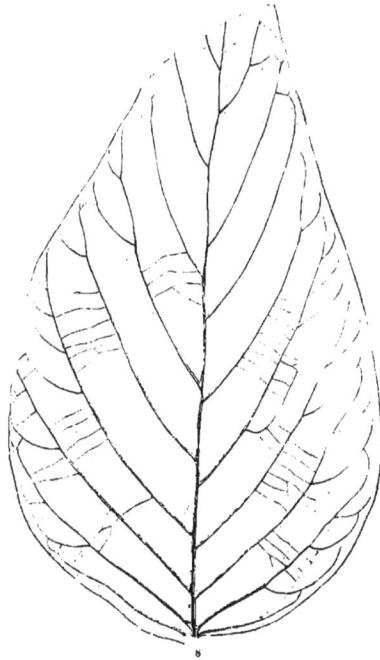

48

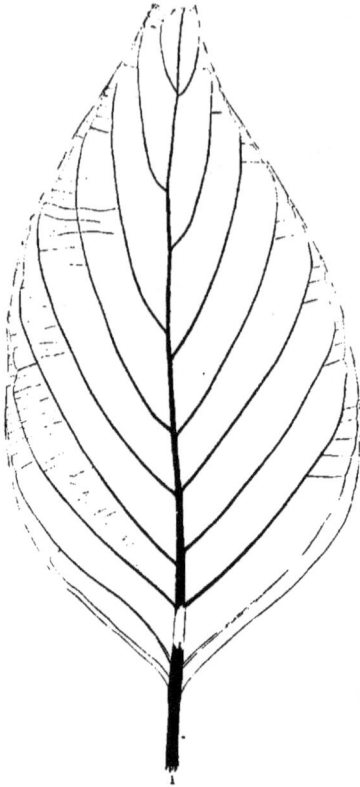

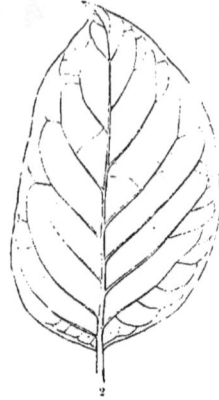

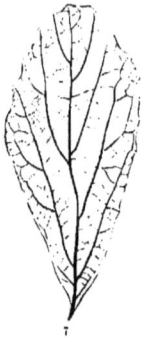

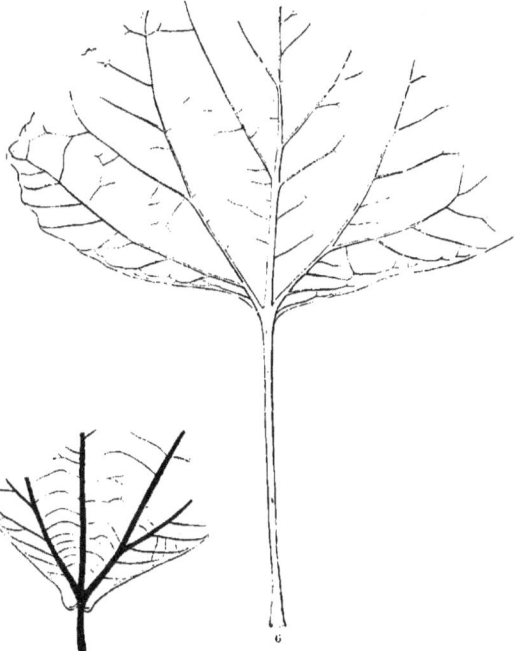

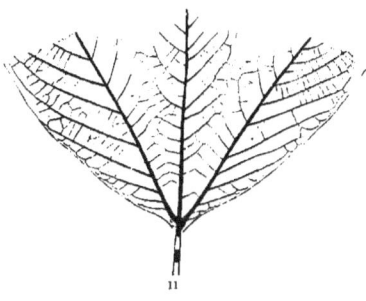

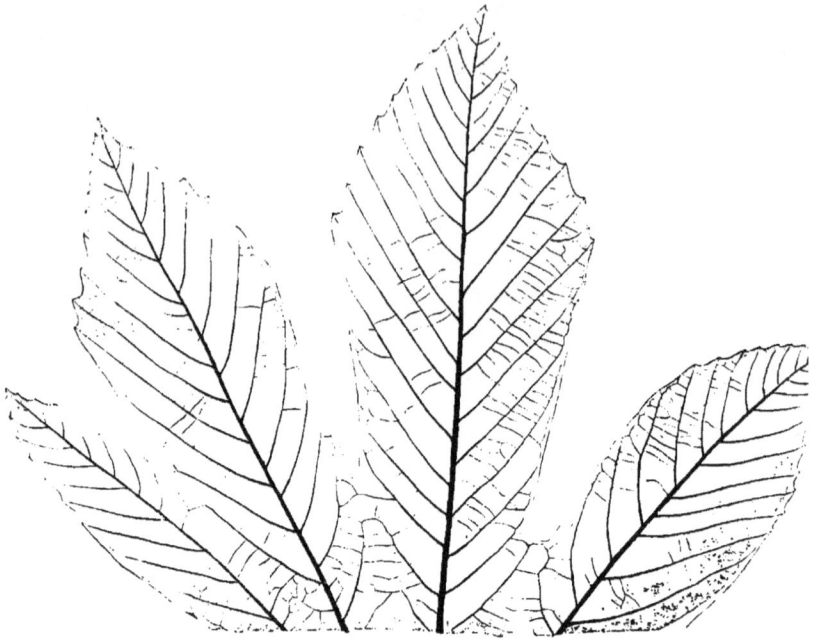

49

1

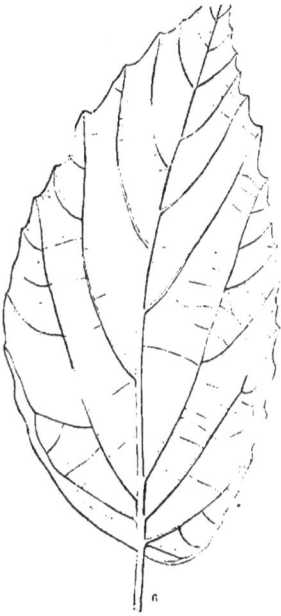

50

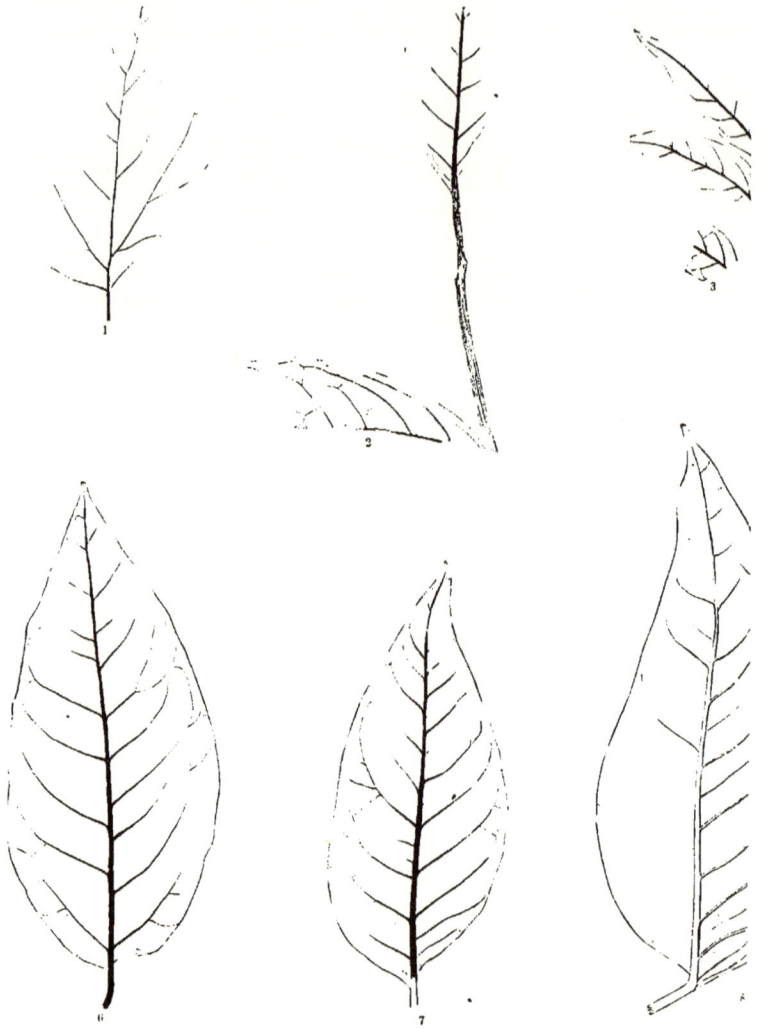

Fig. 1. Acer indivisum. Web.   Figs. 2, 3. Sapindus affinis. Newb.

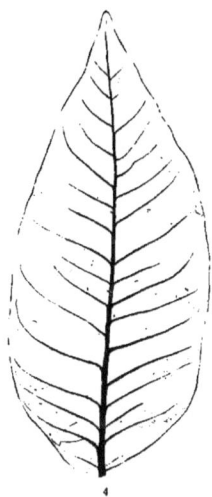

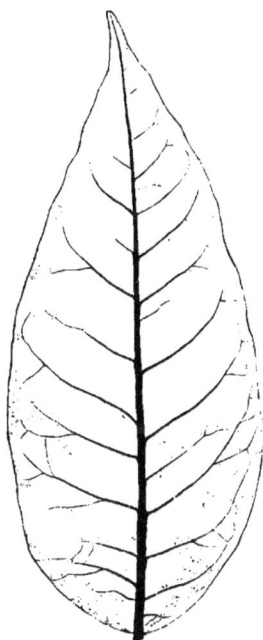

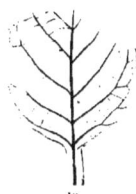

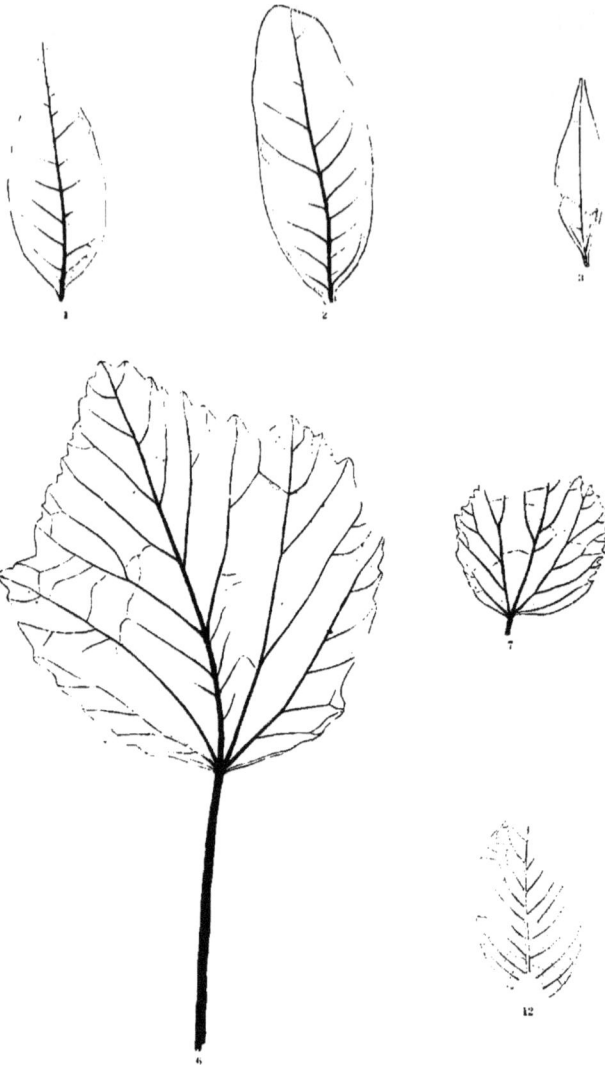

51

FIGS. 1-3. Sapindus angustifolius, Lx.   FIGS. 4-5. Vitis Bruneti, n. sp.   FIG. 6. V. Carbonensis, n.

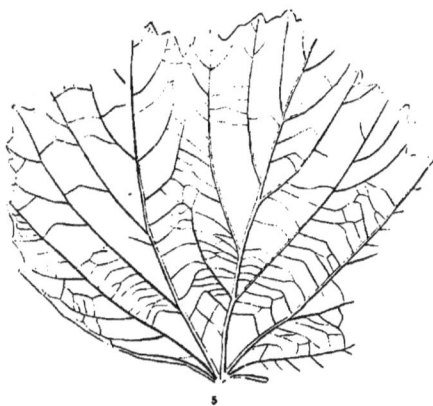

n. sp.   Figs. 9-11. V. cuspidata. n. sp.   Figs. 12, 13. Berchemia multinervis Al. Br.   Figs. 14, 15. Zizyphus serrulata. n. sp.

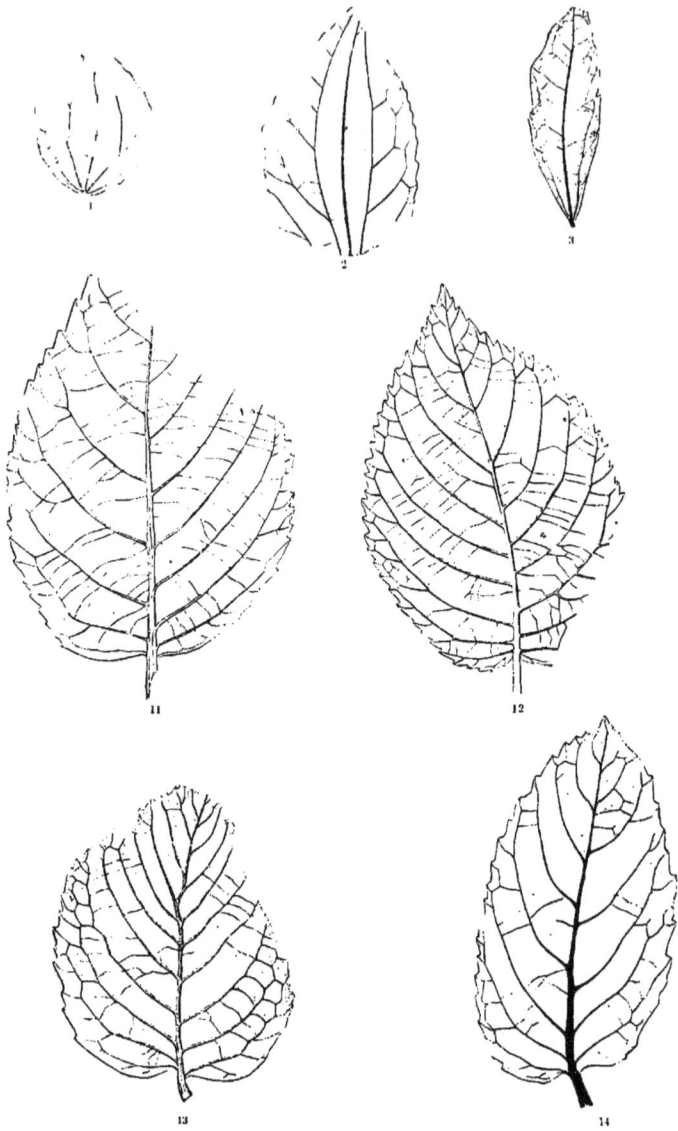

52

Figs. 1, 2. Zizyphus Merkii, Lx.   Fig. 3. Z. cinnamomoides, Lx.   Figs 4-6. Palinrus Colombi. Heer.

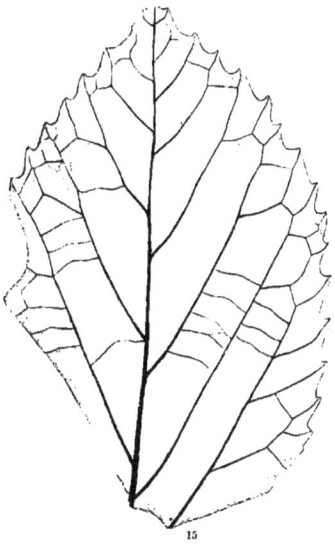

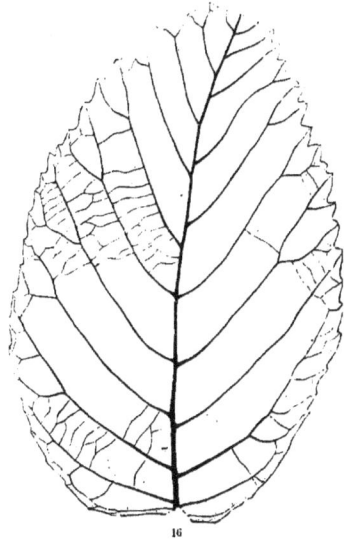

DONS

ima n. sp.   FIGS 8-10. P. Pealei, n. sp.   FIGS. 11-14. Celastrus ferrugineus, n. sp   FIGS. 15, 16. C. Taurinensis, n. sp.

53

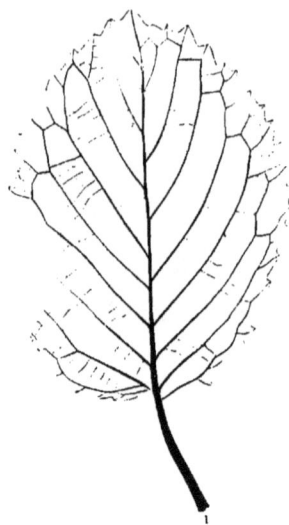

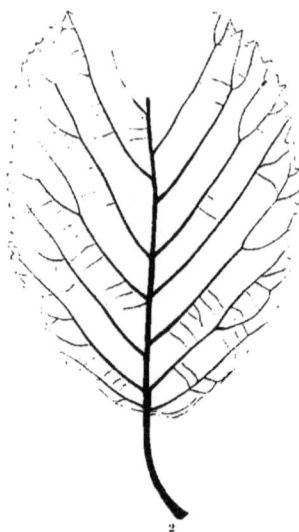

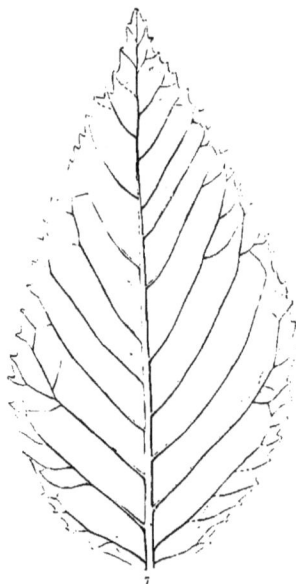

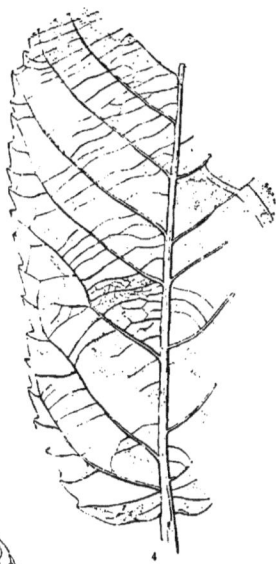

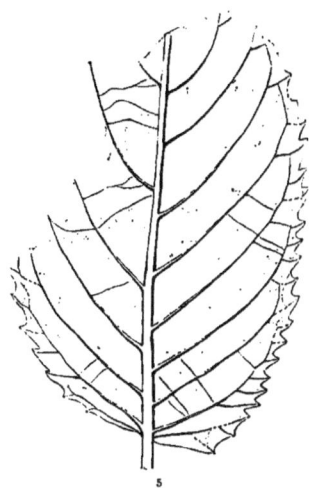

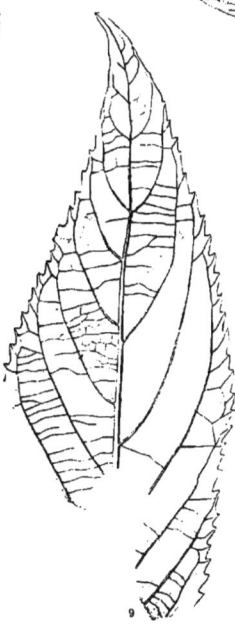

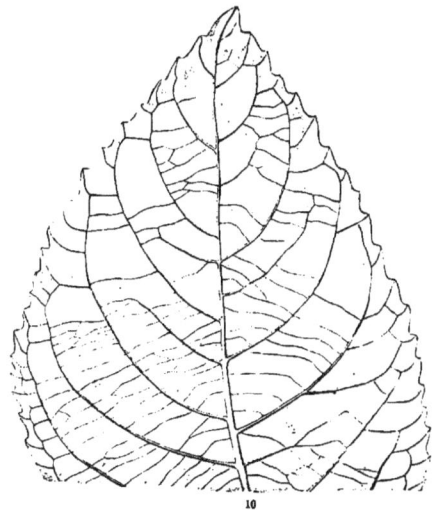

. ovalns. n. sp.　　Fig. 8. C. grewiopsis, n. sp.　　Figs. 9. 10. C. curvinervis, n. sp.

54.

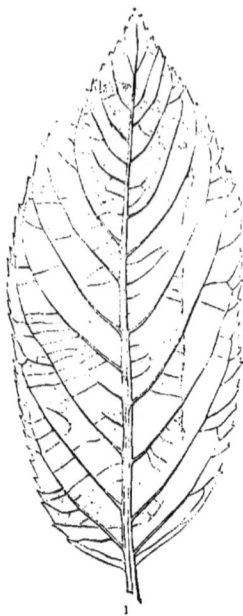

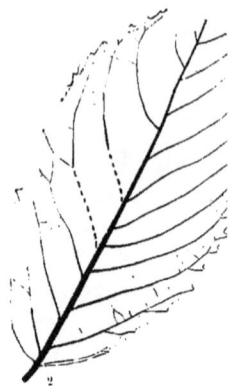

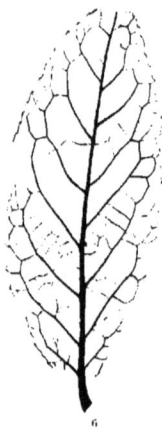

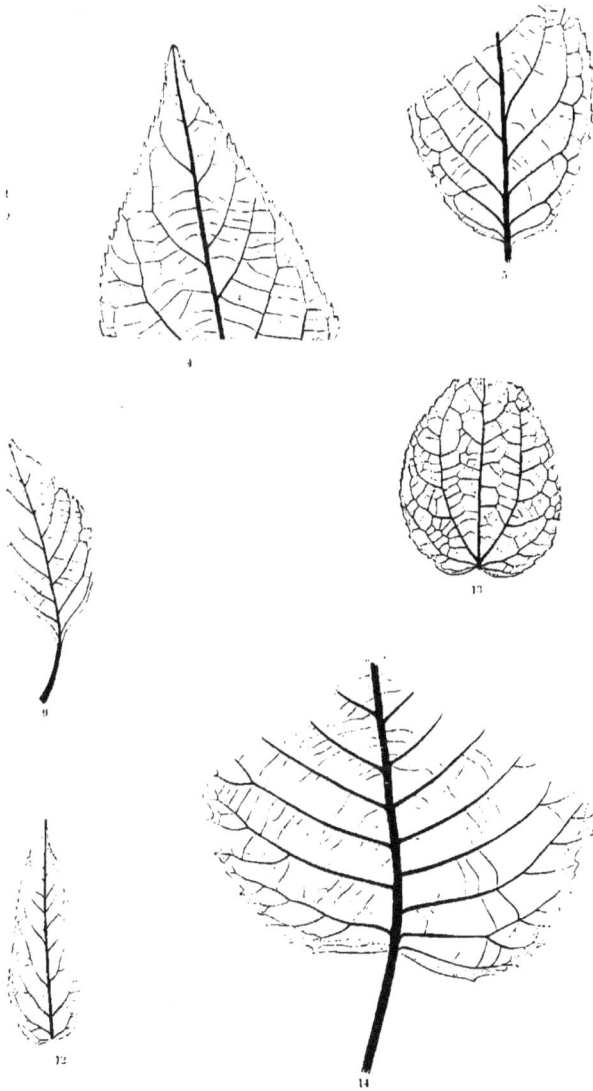

FIG. 13. Grewia crenata (Ung.) Heer.   FIG. 14. G. celastroides, n. sp.

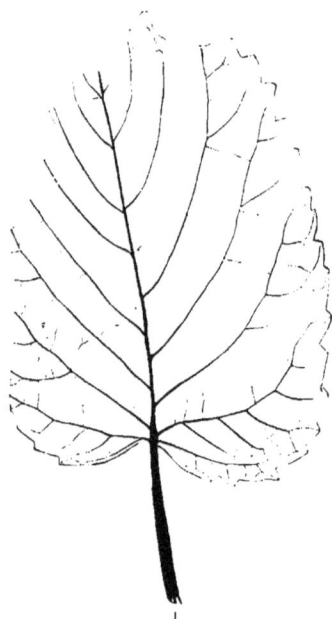

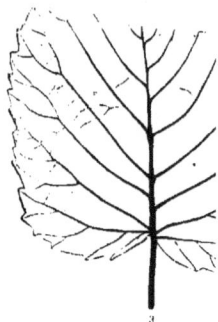

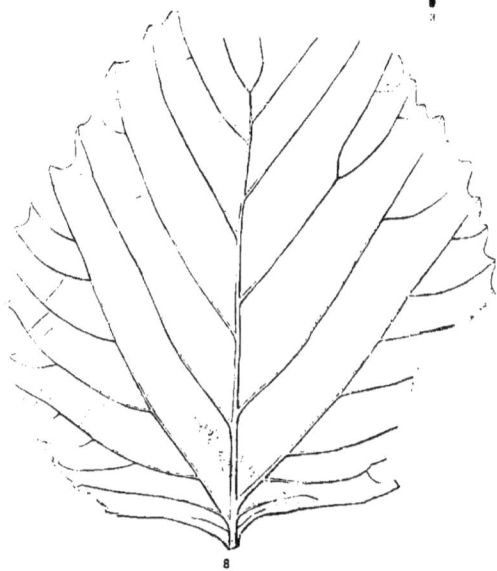

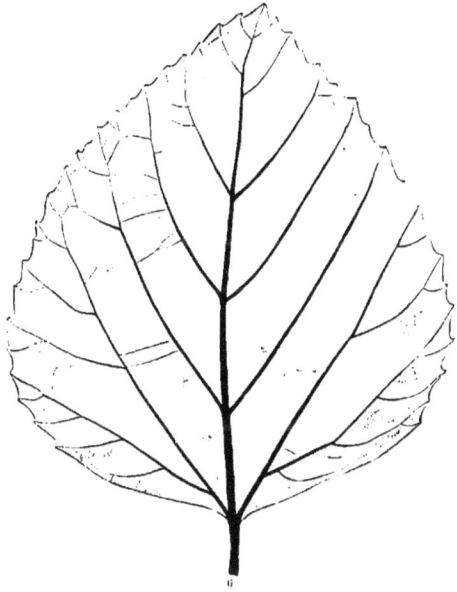

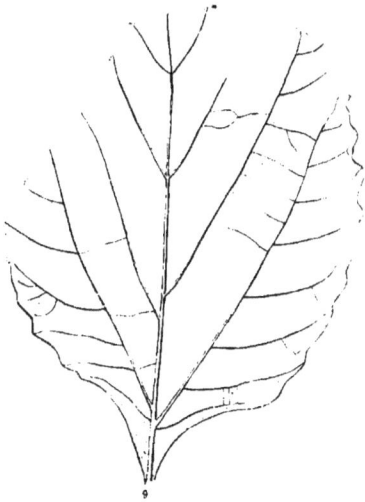

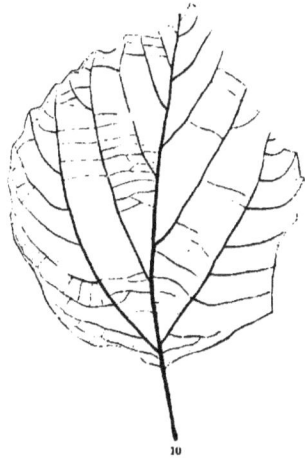

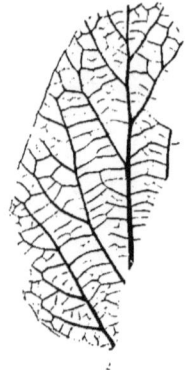

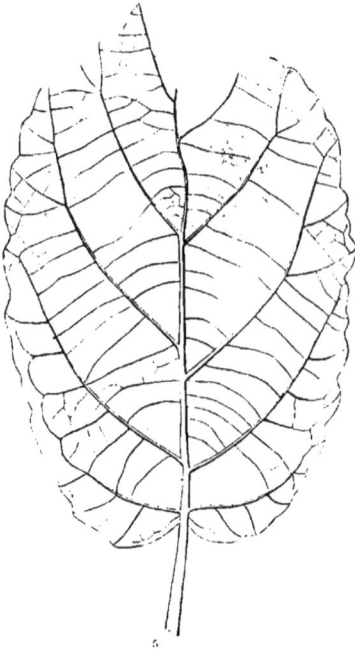

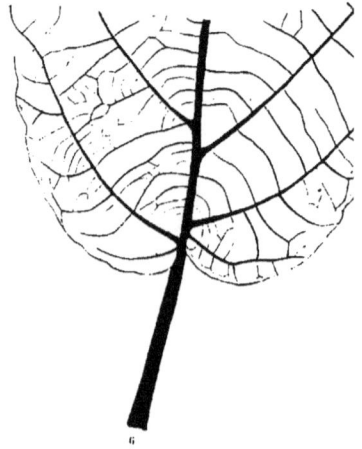

Figs. 1, 2. Grewiopsis ficifolia, n. sp.   Fig. 3. G. paliurifolia, n. sp.   Fig. 4

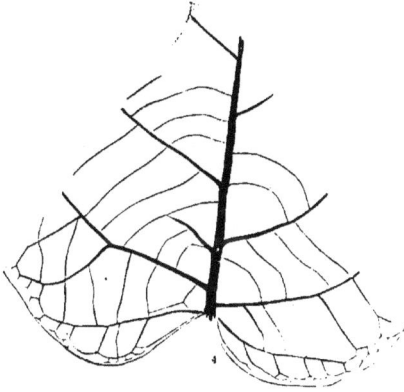

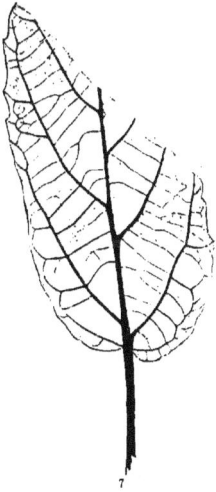

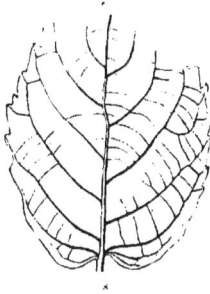

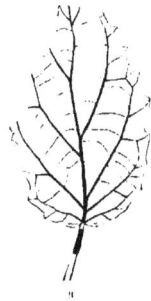

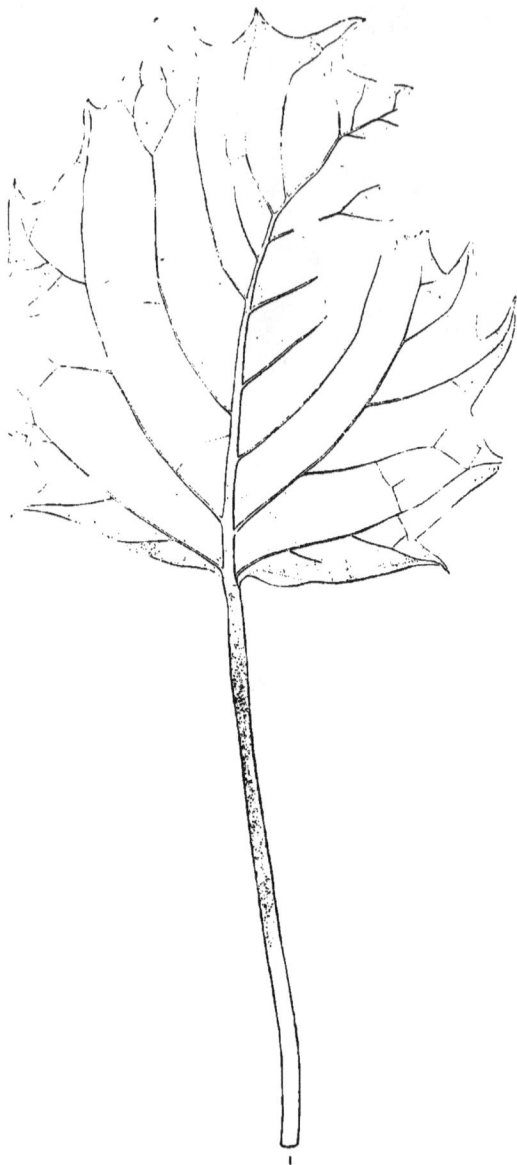

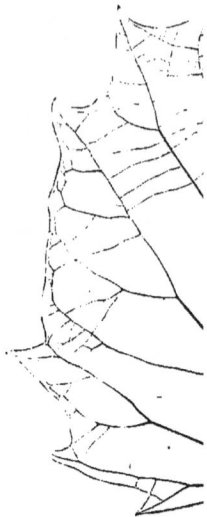

FIGS. 1-5. Ctenidei

58

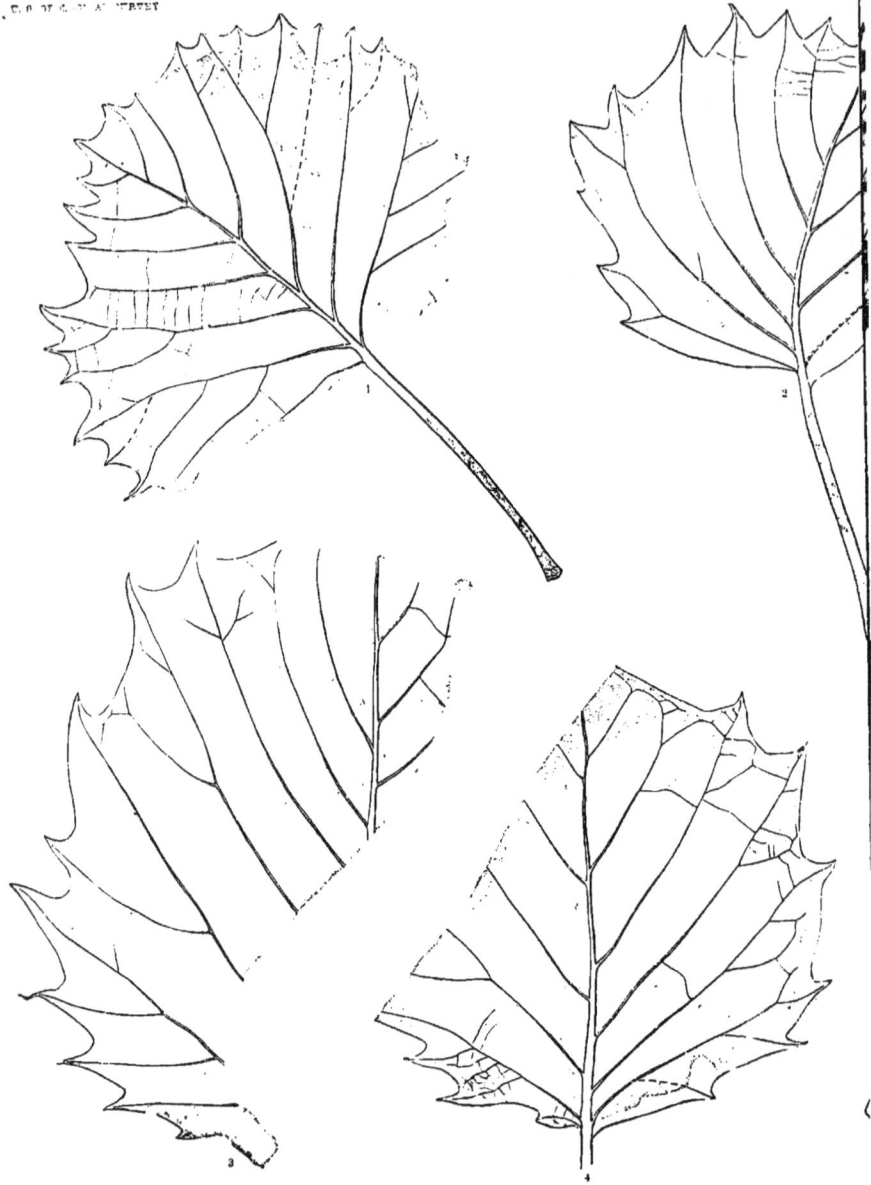

Figs 1-5. Credneria? daturæfolia,

59

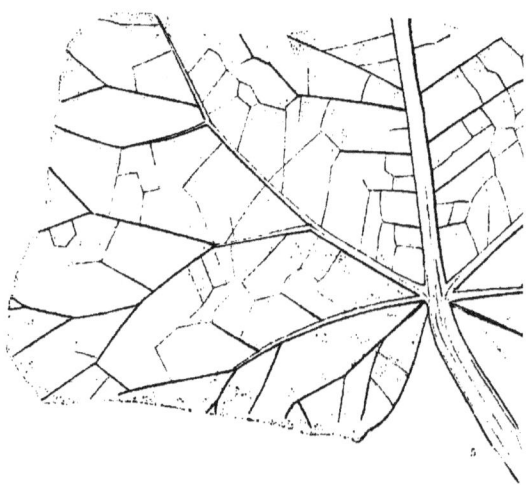

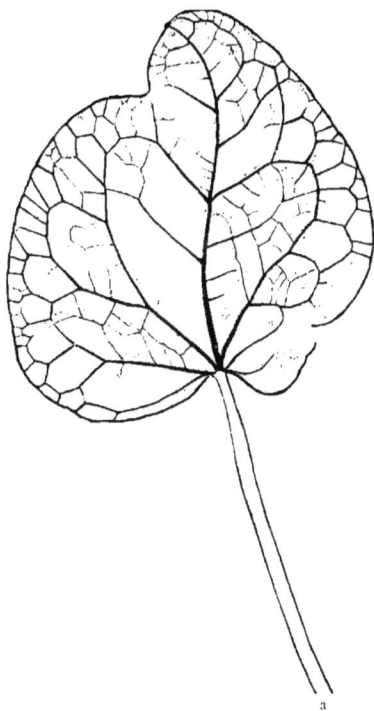

3

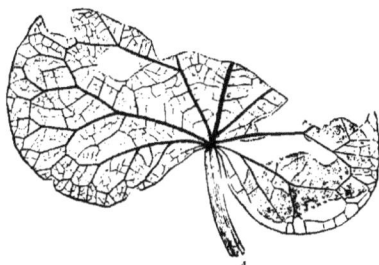

4

60

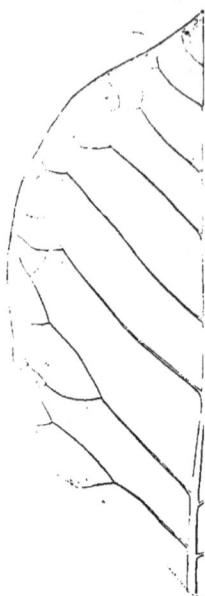

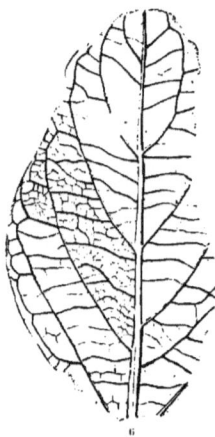

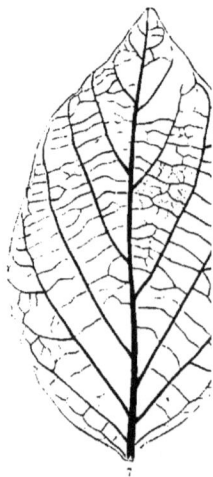

Fig. 1. Liriodendron Laramiense, n. sp.   Figs. 2. 3. Magnolia pulchra, n. sp.

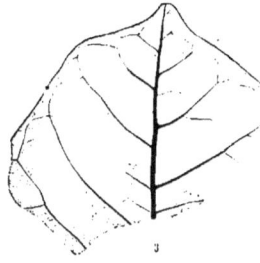

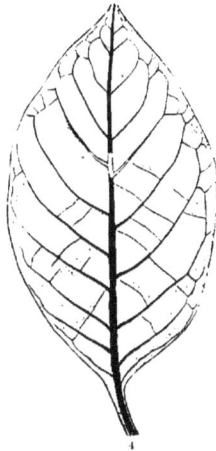

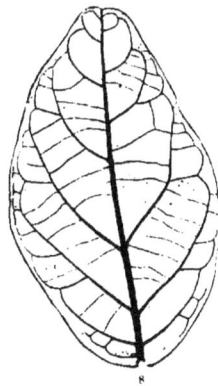

iospyros brachysepala. Al. Br.   Figs. 6, 7. D. ficoidea. Lx.   Fig. 8. D. ? obtusata n. sp

61

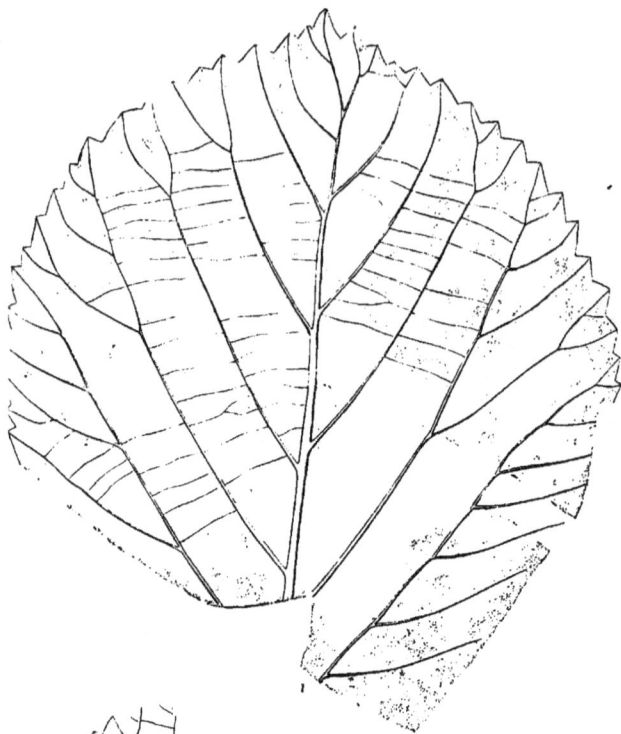

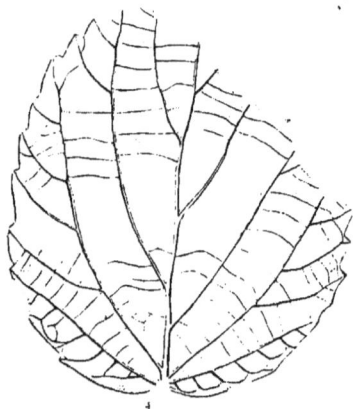

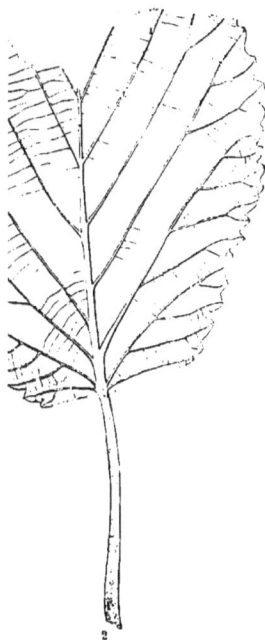

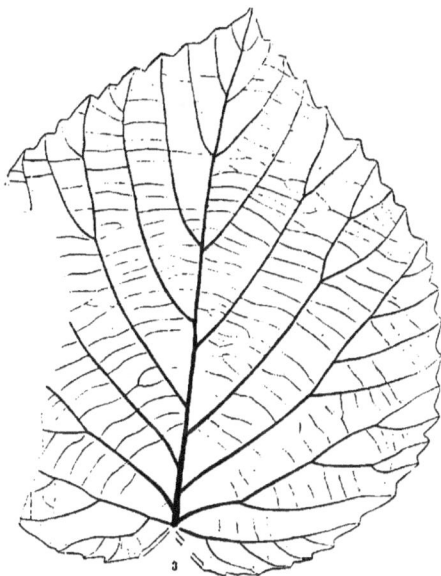

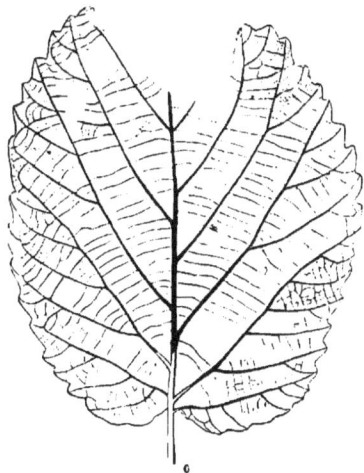

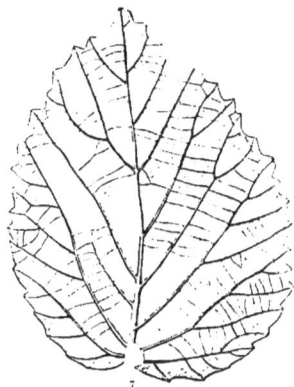

urnum tilioides.

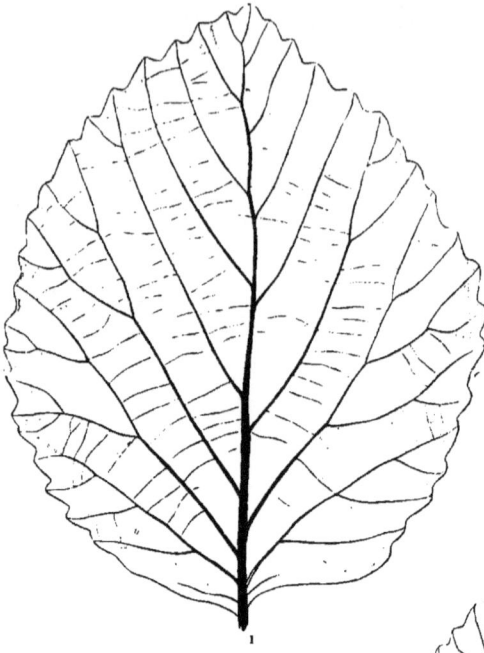

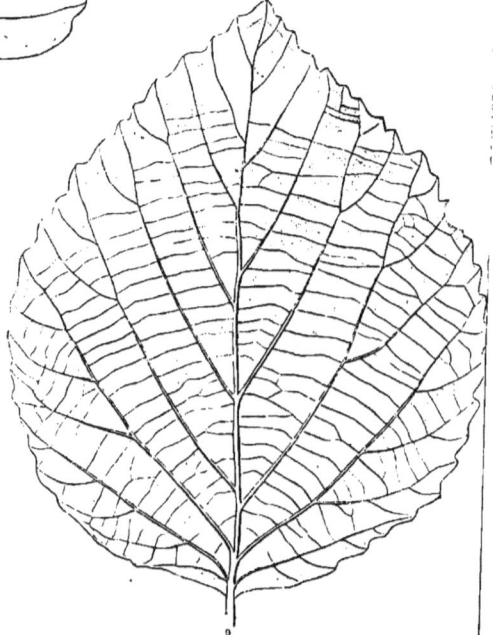

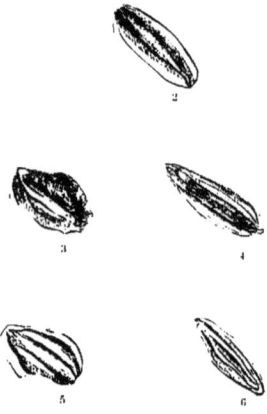

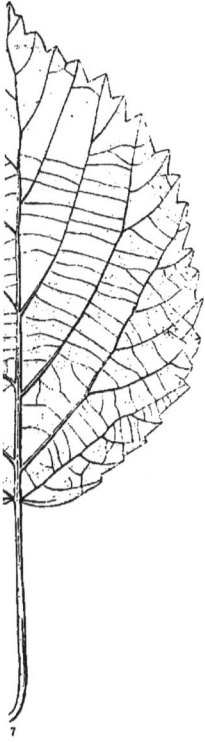

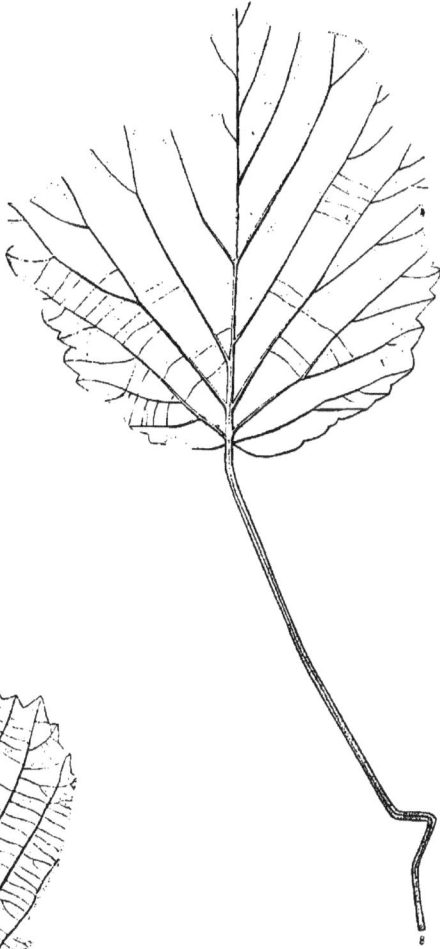

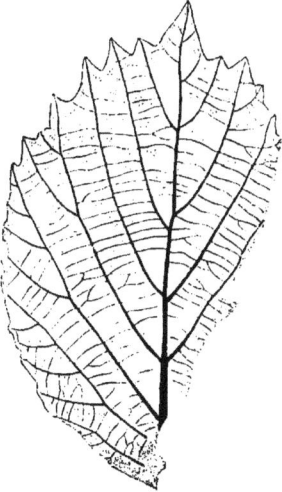

ectum, n. sp.   FIG. 10. V. macrodontum, n. sp.

63.

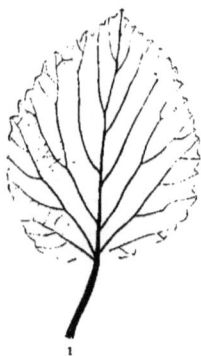

1

2

6

7

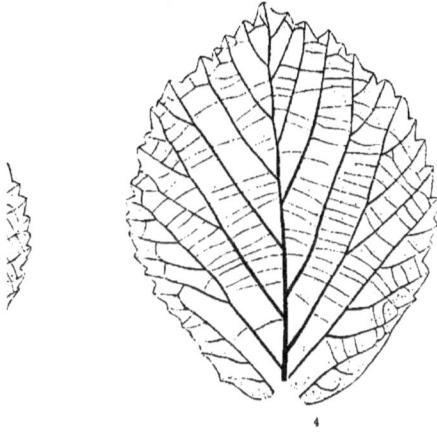

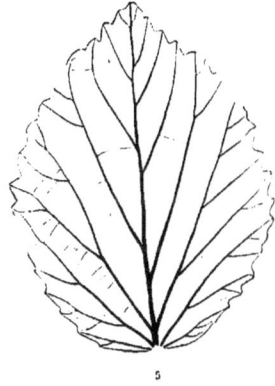

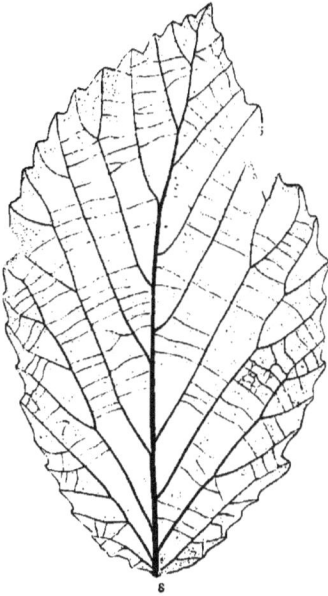

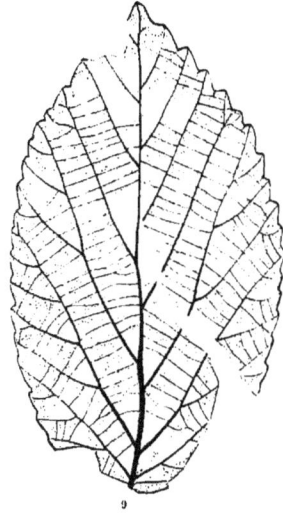

Figs. 6, 7. V. perplexum, n. sp.   Figs. 8, 9. V. elongatum, n. sp.

64

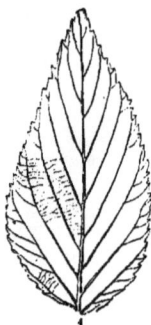

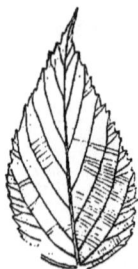

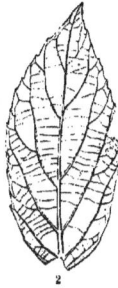

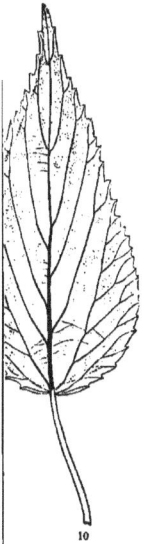

FIGS. 4-9. V. asperum. Newby.   FIGS. 10-12. V. Newberrianum, n. sp.

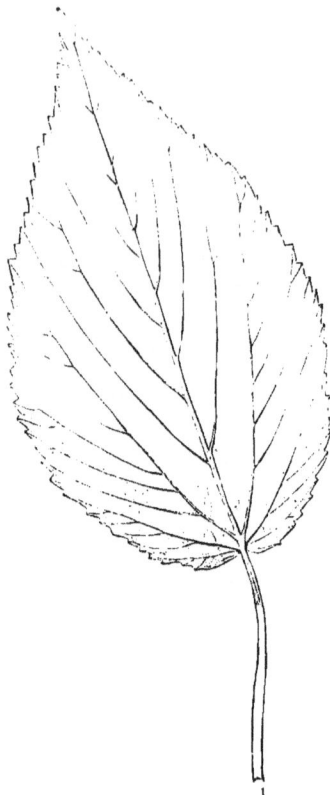

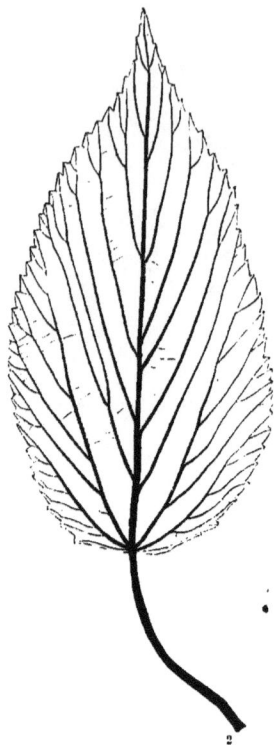

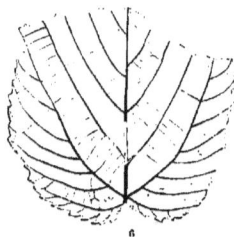

Figs. 1-3. Viburnum Newberrianum, n. sp.   Figs. 4-6. V. Nordens

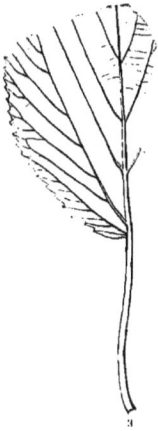

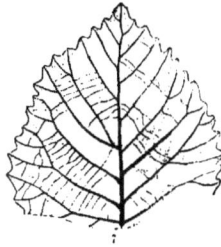

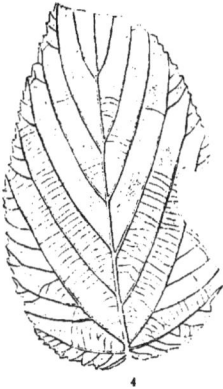

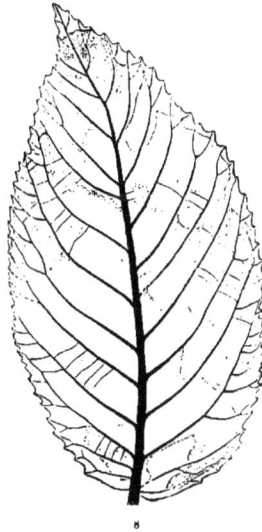

V. betulifolium, n. sp.   FIG 8. V. finale, n. sp.

www.ingramcontent.com/pod-product-compliance
Lightning Source LLC
Chambersburg PA
CBHW021507210326
41599CB00012B/1170